Study of Nuclear and Alternative Energy Systems

SUPPORTING PAPER 6

DOMESTIC POTENTIAL OF SOLAR AND OTHER RENEWABLE ENERGY SOURCES

The Report of the
Solar Resource Group
Supply and Delivery Panel
of the
Committee on Nuclear and Alternative Energy Systems
National Research Council

NATIONAL ACADEMY OF SCIENCES
WASHINGTON, D.C. 1979

International Standard Book Number 0-309-02927-9

Library of Congress Catalog Card Number 79-90523

Available from:

Office of Publications
National Academy of Sciences
2101 Constitution Avenue, N.W.
Washington, D.C. 20418

Printed in the United States of America

PREFACE

In June 1975, the National Research Council (NRC) undertook a comprehensive study of the nation's prospective energy economy during the period 1985-2010, with special attention to the role of nuclear power among the alternative energy systems. The goal of the study is to assist the American people and government in formulating energy policy.

The Governing Board of the National Research Council appointed an NRC-wide Committee on Nuclear and Alternative Energy Systems (CONAES) to conduct the study. CONAES consists of 15 members drawn from diverse disciplines and backgrounds. The committee developed a three-tiered functional structure for the study. The first tier is CONAES itself. The ultimate findings, judgments, and conclusions of the study will be embodied in its final report.

To provide scientific and engineering data and analyses, a second tier of four panels was formed to examine (1) energy demand and conservation, (2) energy supply and delivery systems, (3) risks and impacts of energy supply and use, and (4) syntheses of diverse models of future energy economies, respectively. Each panel, in turn, established a number of resource groups--22 in all--as the third tier, to address in detail an array of more particular matters, such as buildings and transportation systems, solar energy, breeder reactors, coal technologies, health and environmental implications, and alternative consumption patterns and economic models. In all, more than 200 informed individuals served on or contributed to the work of the panels and resource groups.

The National Research Council customarily publishes only the final reports of its committees--and then only after the report has been reviewed by a group other than its authors, according to procedures approved by a Report Review Committee consisting of members of the National Academy of Sciences, the National Academy of Engineering, and the Institute of Medicine. However, because such a large volume of information and analyses was assembled for consideration by the committee, and because of the diversity and scope of that information and the accompanying judgments, the panel reports and approximately 10 reports by the resource groups are being published as supporting papers. Each of these has been considered and used by CONAES but has not undergone the critical review procedure normal to the NRC.

It must be recognized that some conclusions of the panel and resource group reports may be at variance with the conclusions of the CONAES report. The findings reported in these documents are those of their authors and are not necessarily endorsed by CONAES or the National Research Council.

This report covers the work of the Solar Resource Group of the CONAES Supply and Delivery Panel, done for the most part in 1976. The report was intended to inform the panel and CONAES about the current state of solar technology and government policy affecting the use of solar energy. Because the material in this report has not been updated since late 1976, the reader is urged to refer to more recent reports for current information. Among the federal government reports on the subject are:

Office of Technology Assessment, U.S. Congress. 1978. Application of Solar Technology to Today's Energy Needs. 2 vols. Washington, D.C.: U.S. Government Printing Office (052-003-00539-5).

U.S. Council on Environmental Quality. 1978. Solar Energy: Progress and Promise. Washington, D.C.: U.S. Government Printing Office (041-0011-00036-0).

U.S. Department of Energy. 1978. Solar Energy: A Status Report. Washington, D.C.: U.S. Department of Energy (DOE/ET-0062).

U.S. Department of Energy. 1978. Solar Energy Research and Development: Program Balance, A Review by the Solar Working Group. Supporting annex issued as Report No. HCP/M2693-01 and -02. Washington, D.C.: U.S. Department of Energy (IR-0004).

SOLAR RESOURCE GROUP

MELVIN K. SIMMONS (*Chairman*), Assistant Director for Analysis and Assessment, Solar Energy Research Institute

BRUCE ANDERSON, President, Total Environmental Action, Inc.

PIET B. BOS, Department Director, New Energy Resources, Electric Power Research Institute

SHELDON H. BUTT, Market Research & Planning Department, Olin Brass Group, Olin Corporation

WILLIAM L. DICKINSON, Solar Projects Leader, Lawrence Livermore Laboratory

ELMER L. GADEN, Dean, College of Engineering, Mathematics, and Business Administration, University of Vermont

JOSEPH C. GROSSKREUTZ, Assistant Director for Research, Solar Energy Research Institute

RONAL W. LARSON, Solar Energy Research Institute

ABRAHIM LAVI, United States Department of Energy

ADEN B. MEINEL, Professor, Optical Sciences Center, University of Arizona

MARSHAL MERRIAM, Associate Professor, Department of Materials Science, University of California, Berkeley

MICHAEL MULCAHY, Research & Environmental Affairs, Boston Edison Company

ROSALIE RUEGG, Applied Economic Program, Center for Building Technology, National Bureau of Standards

ROGER SCHMIDT, Honeywell Corporation

JOSEPH L. SHAY, Department of Infrared Physics and Electronics, Bell Telephone Laboratories

WILLIAM A. THOMAS, American Bar Foundation

MARTIN WOLF, Department of Electrical Engineering and Science, University of Pennsylvania

CONAES SUPPLY AND DELIVERY PANEL

W. KENNETH DAVIS (*Chairman*), Vice President, Bechtel Power Corporation

FLOYD L. CULLER (*Deputy Chairman*), President, Electric Power Research Institute

DONALD G. ALLEN, President, Yankee Atomic Electric Company

PETER L. AUER, Professor, Mechanical and Aerospace Engineering, Cornell University

JAMES BOYD, Consultant, Washington, D.C.

HERMAN M. DIECKAMP, President, General Public Utilities

DEREK P. GREGORY, Assistant Vice President, Energy Systems Research, Institute of Gas Technology

ROBERT C. GUNNESS, Former Vice Chairman & President, Standard Oil Company of Indiana

JOHN W. LANDIS, Senior Vice President, Stone and Webster Engineering

MILTON LEVENSON, Director, Nuclear Power Division, Electric Power Research Institute

ERIC H. REICHL, President, Conoco Coal Development Company

MELVIN K. SIMMONS, Assistant Director for Analysis and Assessment, Solar Energy Research Institute

MORTON C. SMITH, University of California, Los Alamos Scientific Laboratory

A. C. STANOJEV, Vice President, Ebasco Services Inc.

JOHN O. BERGA, Staff Officer

COMMITTEE ON NUCLEAR AND ALTERNATIVE ENERGY SYSTEMS

HARVEY BROOKS (*Co-Chairman*), Benjamin Peirce Professor of Technology and Public Policy, Aiken Computation Laboratory, Harvard University

EDWARD L. GINZTON (*Co-Chairman*), Chairman of the Board, Varian Associates

KENNETH E. BOULDING, Distinguished Professor of Economics, Institute of Behavioral Science, University of Colorado

ROBERT H. CANNON, JR., Chairman, Division of Engineering and Applied Science, California Institute of Technology

RICHARD R. DOELL, Research Geophysicist, U.S. Geological Survey (Retired 1976)

EDWARD J. GORNOWSKI, Executive Vice President, Exxon Research and Engineering Company

JOHN P. HOLDREN, Professor, Energy and Resources Program, University of California at Berkeley

HENDRIK S. HOUTHAKKER, Henry Lee Professor of Economics, Department of Economics, Harvard University

HENRY I. KOHN, Professor, Radiation Biology, Shields Warren Radiation Laboratory

STANLEY J. LEWAND, Vice President, Public Utilities Division, The Chase Manhattan Bank

LUDWIG F. LISCHER, Vice President of Engineering, Commonwealth Edison Company

JOHN C. NEESS, Professor, Zoology, University of Wisconsin

DAVID ROSE, Professor, Nuclear Engineering, Massachusetts Institute of Technology

DAVID SIVE, Attorney at Law, Winer, Neuberger and Sive

BERNARD I. SPINRAD, Professor, Nuclear Engineering, Radiation Center, Oregon State University

---

JACK M. HOLLANDER, Study Director

JOHN O. BERGA, Deputy Study Director

CONTENTS

TABLES

FIGURES

I. DEVELOPMENT OF THE REPORT

PURPOSE

The Solar Resource Group is a unit of the Supply and Delivery Panel of the Committee on Nuclear and Alternative Energy Systems. It was formed to study the present and potential use in the United States of the renewable energy sources generally included under the term solar energy. In particular, the group was asked to estimate the likely energy costs and potential market penetrations of solar technologies under a variety of assumptions in order to allow for the inclusion of solar energy contributions in national energy models for the United States. These estimates include base scenarios of normal development of the solar market and special scenarios in which national policy dictates more rapid development of solar technologies. The group was not asked to assess the overall feasibility or consequences of these scenarios. Such an assessment, based on information from a variety of sources, is being done elsewhere in the CONAES study.

SCOPE

The group considered technologies which include nearly all means of using solar radiation--those where the initial collection is in a manufactured collector, and those where initial collection is by natural processes. The only major form not included in our review is hydropower, which is the responsibility of another resource group. The technologies have been assembled into seven categories which correspond closely to the organization of the Department of Energy's solar energy program: heating and cooling of buildings, industrial process heat, bioconversion, solar thermal conversion, photovoltaic conversion, wind energy conversion, and ocean thermal energy conversion.

THE SOLAR RESOURCE GROUP

The members of the Solar Resource Group were selected to provide a number of capabilities necessary to fulfill the role of the group in the CONAES study. First, each technology is represented on the group by the appropriate expertise of one or more members. Second, the group contains a balance of individuals from industry, from research institutions, including government laboratories, and from universities. Third, the group includes individuals who have considered the institutional and economic environments in which solar energy must be adopted. Finally, the group includes those who are strong advocates for rapid development of solar energy, and those who are more skeptical of the prospects for immediate success.

PROCEDURE

At the initiation of the work of the Solar Resource Group, individual members were selected to write first drafts of the sections of what is now Chapter IV of this report according to the philosophy and methodology discussed at the beginning of that chapter. These drafts have been debated in meetings of the group, and other members of the group have in some cases developed second drafts from the work of the first author. In addition, members of the group developed draft issue papers for consideration by CONAES. These papers were discussed in the group's meetings and are the basis for the commentaries on solar energy R&D and implementation which form the last two chapters of the report.

INTERACTIONS WITH OTHER ELEMENTS OF THE STUDY

This report is intended to contribute to the broader effort of the CONAES study. Thus interactions of the Solar Resource Group with other resource groups and with the panels of the study have been essential to complete our work and properly design our report. We have also provided other information on solar energy that has been required by other elements of the study.

A direct responsibility of the Solar Resource Group has been to provide information to the Supply and Delivery Panel required for the development of the panel's report to the committee.

There are certain resource groups of the Supply and Delivery Panel whose assignments have significant overlap with the work of the Solar Resource Group. The Electricity Conversion, Transmission, Energy Storage, and Hydroelectric Resource Group shares the responsibility for assessing potential impacts upon the nation's electrical utility networks of development of time-varying or stochastic solar energy sources, or of locally sited solar technologies that require backup from the utility grid. The Non-Energy Materials and Resources Group shares responsibility for assessing the impacts of and the constraints upon an accelerated deployment of solar technologies. The overlap of responsibility with the Resource Group on Financial and Institutional Constraints may be noted from the many discussions of such constraints in this report.

There is a natural overlap of both subject matter and approach with the Consumption, Location, and Occupation Patterns Resource Group of the Synthesis Panel. A number of discussions between members of the two groups have helped to clarify how the adoption of solar energy, especially in dispersed technologies, interacts with the development of modified patterns of housing and lifestyle.

The Solar Resource Group has had a major responsibility for providing data to the Modeling Resource Group of the Synthesis Panel so that it could include solar technologies in its model runs of various future scenarios. The first runs of the Modeling Resource Group, performed before the work of the Solar Resource Group had produced preliminary results, did not include any of the solar technologies. The second and later sets of runs included a surrogate for the full mix of solar electric technologies. Those runs include the option for the year 2000 of central station solar thermal conversion at the capital cost of $1,200 per rated kilowatt (kW) and 0.45 load factor. (For the purposes of the models, this was converted to an equivalent cost of $1,730 per kW for a 65 percent capacity factor.) This provided an upper bound to the cost for electricity in model runs for scenarios with limits on both coal and nuclear electricity. See the report of the Modeling Resource Group (National Research Council, 1978) for details.

II. BACKGROUND

PREVIOUS STUDIES

There have already been many studies performed of the potential energy contributions of solar energy in the nation. Table II.1 summarizes the estimates made by some of these studies. There are other studies in progress now, some others being planned, and more certainly to follow. With so many studies being done, what is the justification for adding yet another? It is not that we have been more thorough, or taken a more detailed look, or a broader look, or that we have been wiser than others. Rather it has been our task to examine these many studies (and other studies of individual technologies) for the committee, and report the conclusions that can be generally agreed upon, and the questions yet unresolved, in a form designed for use in the other efforts of the CONAES study.

It is appropriate that the list of studies in Table II.1 begin with the study by Donovan *et al.* (1972), prepared at the University of Maryland for the National Science Foundation (NSF) and the National Aeronautics and Space Administration (NASA). The widely read report of this study was crucial in awakening interest in solar energy in the early 1970s, interest that had lain dormant since the 1950s--the previous "Golden Age" of solar energy research. While many of the cost estimates of the NSF/NASA study now appear unrealistic, the report did create interest in these technologies and helped initiate the crucial research and development efforts that form the basis of the present solar energy program.

No other single study has had the dramatic impact of the NSF/NASA study. However, more recent reports have gone far beyond that early effort in assessing the functions of our energy system to which solar energy might best contribute, investigating what other energy resources might be partially displaced by solar energy, estimating the cost of the solar energy in delivered and useful form, and assessing what material and human resources will be needed for the implementation of the solar technologies.

Part of the group's task has been to try to unravel the sources of the great variance in the estimates of future contributions of solar energy as shown in Table II.1. We can identify three major sources of these variances.

First are uncertainties about the solar technologies themselves. In only a few cases in these studies has there been significant and relevant experience that provides a solid basis for cost estimates. In most cases we must depend upon design studies that reflect various degrees of optimism about the potential for success in R&D efforts. Perhaps the most drastic example of this is the uncertainty about achieving an advanced technology that provides drastic cost reductions (and reductions in energy inputs) for photovoltaic arrays.

A second source of the variance are uncertainties about the cost and availability of competing energy sources. Studies done by individuals devoted to solar energy sometimes use unrealistically high estimates for the future prices of fuels; assumptions, for instance, of rapid exponential growth in the cost of fossil fuels are not uncommon. Studies based on such assumptions tend to be unrealistic to the degree that they fail to account for the drastic primary and secondary impacts on our economy of fuel prices many times their present values. Within the CONAES study this uncertainty is at least made clear since the assumptions about other fuels and their prices are being developed as separate efforts, and efforts are being made in developing the future scenarios of the energy system to include the impacts of high energy prices on the structure of the economy and its ability to provide new investment capital.

The third source of the variance are uncertainties about the financial and institutional environment in which solar energy will be used. The lack of a national energy policy, and in particular lack of a solar energy policy, have contributed to this uncertainty. Though all conventional sources of energy enjoy certain preferential treatment in tax policy and have other subsidies, many studies have assumed that solar energy will be provided no such advantages. Because solar energy is intensive in first cost, the charge on capital for solar devices is an extremely important parameter in determining solar energy's ability to compete in the market. Policies that provide for low interest rates for solar devices would be as important to the development of solar energy as the oil depletion allowance has been to the petroleum industry. The recent moves in Congress to establish tax exemptions for solar energy systems for buildings may define much better the assumptions that should be used regarding the financial environment in which solar energy will compete.

SPECIAL CHARACTERISTICS OF SOLAR ENERGY

Solar energy differs in many respects from other more conventional forms; these differences have important implications for the manner in which solar energy is evaluated and integrated into our society.

(1) Unlike most other forms of energy, solar energy is relatively well distributed across the nation, and is available in relatively low concentrations. This implies a natural bias toward localized, onsite collection, storage, and distribution systems; however, other factors, such as market distribution costs, may partially offset this bias.

(2) Although solar energy conversion is resource intensive due to the fact that large areas of collection surface are required, most resources needed for solar energy technologies are relatively abundant. Thus there is generally not any resource restraint on the amount of solar energy that can be captured and used.

TABLE II.1 Forecasts of Solar Energy Applications in the Year 2000, in Quads

Source	Date	Solar Thermal & Photovoltaic	Wind & OTEC[a]	Other Solar	Total
Donovan *et al.*[b]	Dec 72	3.1	2.4	6.5	12.0
NASA/ASEE[c]	Sept 73	0	0	3.5-6.1	3.5-6.1
Solar Hearings[d]	Nov 73	5.3	15	4.5	24.8
MITRE[e]	Nov 73	7-18	11-27	18	36-63
CEQ[f]	Mar 74	0	0	1	1
Ford Found.[g]	Late 74	1	1	2	4
FEA[h]	Nov 74	6-27	19-40	8-27	34-94
Teller[i]	Apr 75	0	0	5	5
ERDA-48[j]	June 75	1.3-6.1	0	5	6.3-11.1
ERDA-49[k]	June 75	3.1	1.8	6	10.9
Candela and Wiener[l]	Aug 75	6-9	6-9	6-9	18-27

[a] Ocean Thermal Energy Conversion.

[b] Donovan, P. I., W. Woodward, W. R. Cherry, F. A. Morse, and L. O. Herwig. 1972. An Assessment of Solar Energy as A National Energy Resource. Springfield, Va.: National Technical Information Service (PB-221659).

[c] National Aeronautics and Space Administration and American Society of Electrical Engineers System Design Summer Faculty Program. 1973. TERRASTAR: Terrestrial Applications of Solar Technology and Research, Final Report. Springfield, Va.: National Technical Information Service (N-74-12674).

[d] U.S. House of Representatives. 1974. Hearings on the Solar Energy Research, Development, and Demonstration Act, H.R. 10952. Subcommittee on Energy, Committee on Science and Astronautics. 93rd Cong., 1st sess. Washington, D.C.: U.S. Government Printing Office.

[e] MITRE Corporation. 1973. Systems Analysis of Solar Energy Programs. McLean, Va.: MITRE Corporation (MTR-6513).

[f] U.S. Council on Environmental Quality. 1974. A National Energy Conservation Program: The Half and Half Plan. Washington, D.C.: U.S. Government Printing Office.

[g] Ford Foundation. 1974. A Time to Choose: America's Energy Future. Final Report by the Energy Policy Project of the Ford Foundation. Cambridge, Mass.: Ballinger.

[h] Federal Energy Administration. 1974. Project Independence Blueprint, Final Task Force Report: Solar Energy. Springfield, Va.: National Technical Information Service (PB-248507).

[i] Teller, E. 1975. Energy: A Plan for Action. Report to the Energy Panel of The Commission on Critical Choices for Americans. New York: The Third Century Corporation.

[j] U.S. Energy Research and Development Administration. 1975a. A National Plan for Energy Research, Development and Demonstration. Washington, D.C.: U.S. Energy Research and Development Administration (ERDA 48).

[k] U.S. Energy Research and Development Administration. 1975b. National Solar Energy Research, Development and Demonstration Program: Definition Report. Washington, D.C.: U.S. Energy Research and Development Administration (ERDA 49).

[l] Candela, B. J., and A. J. Wiener. 1975. Lifestyle and Energy Consumption: Scenarios for the Years 1985, 2000 and 2025--Alternative Coal Futures in the American Economy. Task 5 of Issues Relative to the Development and Commercialization of a Coal-Derived Synthetic Liquids Industry. Springfield, Va.: National Technical Information Service (FE-1752-1).

(3) Although no technology is completely free of environmental and public health impacts, solar energy technologies appear, on the basis of our present knowledge, generally more benign than the conventional sources of energy. The one exception may be certain forms of bioconversion whose pollutant emissions may be comparable to those from combustion of fossil fuels.

(4) Public attitudes toward solar energy development are almost without exception positive. This is in striking contrast to other energy technologies, where public opposition to some new facilities must be a major consideration in planning. However, the attitude of environmental groups has shifted somewhat in the last few years from full support of all solar technologies. There is now a growing trend to question the wisdom of large, centralized solar energy systems and to prefer local, decentralized systems.

(5) Solar energy systems tend to attain optimum performance in modules that are smaller in scale than those common in other energy technologies (e.g., coal or nuclear electrical generating plants, or the proposed coal conversion plants). As a consequence, the lead time necessary to begin operating a solar energy facility can be much shorter than for other types of energy production facilities.

(6) The use of solar energy for applications such as space and water heating is relatively simple, easy to understand, and implement. Therefore, many individuals with adequate training and instruction will be capable of acquiring and using the simpler and less complex solar technologies.

(7) Distributed solar energy systems, if not properly designed, can have an adverse impact on the operating efficiency of utilities by worsening the ratio of peak-to-average load for a utility. This would worsen the economics of operation of the utility, and lead to a requirement for excessive amounts of installed generating capacity. On the other hand, properly designed control systems allow such solar installations to use their storage capacity to improve the peak-to-average load ratio and make the use of the utilities' generating capacity and distribution system more, not less, efficient.

THE SIZE OF THE SOLAR RESOURCE

The size of available domestic and world resources has been an important consideration for other resource groups of the Supply and Delivery Panel, but not for the Solar Resource Group. Simply put, the solar resource is huge. The important questions have to do not with exact definitions of its size, but with how we might utilize even a small fraction of it.

Solar energy arrives at the surface of the United States at an average rate (day and night, year-round) of 4.76 kilowatt-hours (kWh) per square meter per day (Donovan *et al.*, 1972), so over the course of a year, one square kilometer would receive 1.7 billion kWh. In 1974, the total energy consumed by the United States for all purposes was about 24 trillion kWh. Accordingly, 141,000 square kilometers of land, dedicated to solar energy conversion at 10 percent efficiency, could meet, on the average, the entire United States energy requirement.

The domestic United States wind energy resource has been estimated at about 100 billion kW (Donovan *et al.* 1972), greatly exceeding our present electrical generating capacity. The resource available for ocean thermal energy conversion similarly exceeds our present total electrical generating capacity.

III. SCENARIOS FOR THE SUPPLY AND DELIVERY PANEL

The Solar Resource Group has not developed a set of full scenarios for the various possible levels of future use of solar energy. The considerations that enter into the development of such scenarios include many factors beyond the assignment of the Solar Resource Group. For discussion of these factors, see the reports of the Modeling Resource Group (National Research Council, 1978), the Consumption, Location, and Occupational Patterns Resource Group (1977), the Decision Making Resource Group (1977), and the Demand and Conservation Panel (National Research Council, 1979).

However, the Solar Resource Group has been requested to provide the Supply and Delivery Panel with sets of values for energy production by solar and other renewable resources that might be usefully considered in the context of corresponding production scenarios for conventional energy technologies and for the breeder reactor. Accordingly, presented here is a set of production schedules based upon selected values drawn from the work of the Solar Resource Group and other groups of CONAES. These scenarios are presented only as examples of the kind of production estimates that can be developed under different assumptions, and do not represent a consensus of the members of the Solar Resource Group.

HIGH SOLAR ENERGY SCENARIO

This scenario is based on the assumption that by 1985 a national policy decision will have been made to apply vigorous means to bring about the use of solar energy in each of its various applications. This is a scenario driven not by normal economic forces but rather by a policy of intervention in the energy market. It is assumed that this policy will require, after 1990, adoption of solar energy for all new buildings and for all industrial process heat for which it is applicable; will require rapid adoption of technologies for conversion of municipal and agricultural wastes for energy; and will set in motion a schedule of deployment of several solar electric technologies.

Direct Use of Solar Energy

We assume that all new construction after 1990 will use solar space heating, and that half of the heat collected for this purpose will also be used for solar total energy systems. The accelerated schedule for use of industrial process heat in the Solar Resource Group report is assumed (see Chapter IV), and it is again assumed that half of that heat will be used in solar total energy systems. With these assumptions, the schedule of energy contributions in quads for direct use of solar energy is as follows (not including the electricity produced in the total energy systems):

	1985	1990	2000	2010
Domestic Water Heating	0.2	0.4	1.2	1.3
Passive Space Heating	0.1	0.2	0.3	0.4
Active Space Heating	0.02	0.06	0.64	1.21
Nonresidential Air Cond.	--	0.06	0.36	1.50
Industrial Process Heat	0.2	0.4	1.6	6.6
	0.5	1.1	4.1	11.0

Biomass Conversion

We assume that a decision will be made to rapidly develop systems to recover the energy content of municipal waste and agricultural residues. It is assumed that by the year 2000 95 percent of municipal waste will be processed for its energy, and that 35 percent of the energy content of agricultural residues will be recovered as methane. We do not include here the possible contributions of energy farms. It was estimated in the report of the Solar Resource Group that such farms could provide another 3.4 quads in the year 2010. An option in this scenario is to include that extra contribution.

	1985	1990	2000	2010
Municipal Wastes	0.5	0.75	1.9	1.9
Agricultural Residues	0.5	0.9	3.5	3.5
Energy Farms	--	--	--	--
	1.0	1.65	5.4	5.4

Solar Electric Conversion

There are a number of possible solar electric technologies, and it is not now possible to predict which would be selected for use in a solar-intensive scenario. We base this scenario on the use of three--central station solar thermal conversion, solar thermal total energy systems, and dispersed wind energy systems--without any implication that these would necessarily be the specific technologies used in such a scenario. Photovoltaic or ocean thermal conversion may serve instead of those selected here. For the purposes of this scenario it is assumed that public policy will require the following specified amounts of solar electric capacity for new generating capacity. The estimates for solar thermal central stations are based upon an assumption that in 1990 three full-scale production facilities for heliostats will be placed in operation, and that the production of heliostats will be doubled every 5 years thereafter--a 15 percent annual growth rate. The estimates for solar energy total energy systems are based upon the estimates given above for solar heat delivered for space heating, air conditioning, and industrial process heat. It is assumed that 50 percent of the heat delivered for space heating and processes will be used for total energy conversion, and 70 percent of the heat for air conditioning will be so used. The heat rate of the total energy systems is assumed to be 17,000 Btu per kWh, and the credit for energy displaced is taken at 10,000 Btu per kWh. The estimate for dispersed wind energy systems is taken from the report of the Solar Resource Group. This value is much less than the estimates made by Coty and Dubey (1976) for an accelerated deployment of wind energy: they estimate about 10 times more wind energy capacity in 1995 than is used here. The energy input displaced is then, in quads.

	1985	1990	2000	2010
Central Station	--	--	1.7	8.7
Total Energy	--	0.07	0.5	1.9
Wind	0.1	0.5	1.4	1.8
	0.1	0.6	3.6	12.4

This corresponds to an installed capacity in the year 2010 of 250 GWe of central station solar thermal plants (load factor of 0.4), 74 GWe of total energy generation (load factor of 0.3), and 50 GWe of wind turbines (load factor 0.4).

Combining these estimates, we have the following totals for quads of primary energy:

	1985	1990	2000	2010
Direct Use	0.5	1.1	4.1	11.0
Biomass	1.0	1.6	5.4	5.4
Solar Electricity	0.1	0.6	3.6	12.4
	1.6	3.3	13.1	28.8

LOW SOLAR ENERGY SCENARIO

This scenario is based on the assumption that no policy will be implemented to assist the entry of solar energy into the market, that the costs of other energy sources follow the baseline schedule of the Modeling Resource Group, and that the costs of solar energy technologies follow the estimates of the Solar Resource Group (without breakthrough in advanced technologies). Under these assumptions, estimates for the contributions of direct use of solar energy and for solar electricity can be drawn from other reports of the CONAES study. The Demand and Conservation Panel, in its scenario C, examines the case in which energy costs remain constant in 1975 dollars, except for natural gas, which is expected to nearly quadruple between 1975 and 2010. In this scenario, they find that the direct use of solar energy remains very small, reaching only 0.3 quads in the year 2010. In scenarios with similar assumptions of the future cost of coal and nuclear electricity, the Modeling Resource Group finds no market penetration for solar electricity by the year 2010. Similar estimates of the use of biomass (including waste) for energy have not been made by these groups, but we can take a representative value from the analysis of methane production from municipal solid waste that was performed by Dynatech Corporation (Kispert *et al.*, 1976). Dynatech estimates that about 200 plants might be built to serve the 65 Standard Metropolitan Statistical Areas in the U.S. that have populations over 500,000. These plants would each process about 1,000 tons per day of solid waste and convert the biomass content to methane. These plants would produce a total of 750 million cubic feet per day of methane, about 1.5 percent of present U.S. consumption.

Combining these estimates, we have the following low scenarios for use of solar energy, in quads of primary energy displaced.

	1985	1990	2000	2010
Direct Use	0	0	0	0.3
Biomass	0	0	0.1	0.3
Solar Electricity	0	0	0	0
	0	0	0.1	0.6

IV. SCHEDULES OF ENERGY CONTRIBUTIONS AND COSTS

We report here the best estimates that the Solar Resource Group is now able to make on the possible energy contributions and the probable costs for energy from the major solar technologies. These values must not be taken as predictions; they are not. Nor can they be understood without keeping in mind the procedures we have used in their development. First, the basis we have used in developing cost estimates: we have tried to estimate the cost at which each of these technologies might be available today (1975 cost) and the costs for future years (1985, 1990, 2000, and 2010) under certain assumptions discussed below. Constant 1975 dollars are used throughout, unless otherwise noted. There are uncertainties in the estimates for each of the technologies, but the types and sources of uncertainty are very different. For example, in the case of photovoltaic conversion the present costs can be well specified because some units are now marketed, but this well-defined cost is of no use in trying to estimate the anticipated cost for widespread utilization of photovoltaics in the future. In the case of ocean thermal energy conversion, there is no basis for the estimate of present cost except for limited studies of possible designs. Thus the cost estimate for ocean thermal in 1975 is more uncertain than is the estimate for the 1975 cost of photovoltaics. However, because the advances sought in ocean thermal technology are less revolutionary than those sought in photovoltaic technology, the cost estimate for ocean thermal in the year 2000 is more certain than the year 2000 estimate for photovoltaics.

A learning-curve effect is nearly always assumed by those in the solar energy community when trying to predict the future costs of solar energy devices. It must be admitted that thus far there is little or no evidence of this effect in solar technologies. There has been no reduction in the cost of flat-plate solar collectors; indeed, the prices may be increasing at or faster than the average rate of inflation. However, the reliability and efficiency of flat-plate collectors has improved during this time so their value has increased as well. There have been some significant reductions in the costs of photovoltaic arrays as the industry has started production for terrestrial applications, but it is not known yet if costs will continue to decrease as the volume of production and the experience of the manufacturers increases. It is certainly true that the costs of solar components are generally much higher now than the equivalent prices of similar nonsolar mass-produced items (for example, on the basis of dollars per pound) and this does indicate a significant potential for reduction of cost. In our estimates of costs we have generally tried to use moderate assumptions about the gains to be made from learning curves. It should be understood that if sales lag far behind those estimated, then the experience acquired may be insufficient to allow the estimated costs to be achieved, so that these two types of estimates are strongly coupled. In the case of photovoltaic conversion, the assumptions made about the learning curve are crucial, and the section on photovoltaics should be read carefully to appreciate the meaning of the estimates provided there.

In developing our estimates of the energy contributions of the solar energy technologies, we have assumed the action of market forces in an economic environment in which the technology under consideration is competitive. That is, our estimate is of the sales of solar systems that will occur in a given year if the costs of the competing fuels match the estimated cost of the solar technology for that year. If this competitive situation continues to hold throughout each year of the period under consideration, the market for solar energy should follow approximately the estimated schedule. It is highly unlikely that these costs will actually remain equal in a free market, because under our assumptions the price of the solar technologies generally decreases with time (and experience) and it is probable that the cost of competing fuels will increase with time. Therefore, the relative competitive position of solar technologies will be a strong function of time, with the solar options being generally more expensive in the near term.

It can be seen that the combination of the assumption of a learning-curve effect dependent upon experience with solar technology and a situation of declining relative costs of solar versus nonsolar technologies creates a paradox. Only as solar devices are made and used will experience be obtained, and only if experience is obtained may their prices later become competitive with the nonsolar alternatives. It is probable that government action, in the form of tax exemptions, low-interest loan guarantees, or other incentives, will be required to break through this barrier of inexperience with solar devices and open up a competitive market for solar energy. If such action is forthcoming, and if it is effective, then the energy contributions we have estimated can come about.

It should be noted that some of the options have much higher estimated costs than others. It may well be that some of the more expensive options may not be capable of achieving a competitive position, no matter what subsidies are given. Thus these technologies may not ever be used except in a scenario in which other energy options vanish and we must depend upon solar energy for nearly all of our needs.

It is of course unrealistic to quote only a single price for the energy from a solar technology: the cost of delivered energy will vary according to the characteristics of the location and the demands to be served. It is also false that a technology will have no share at all of the market if its cost is above its competitors, and will dominate the market if its cost is just below. Instead, the share of the market taken by a given solar technology will vary smoothly as a function of the cost of the competitive sources of energy. However, a simple and useful representation of this function is a step of a height corresponding to the anticipated market share of solar energy at equal cost, and placed on a cost scale at the position of equal cost per unit of delivered energy.

Such step functions are shown in Figure IV.1 for our estimates, described in this chapter, of the contributions of solar energy as heat and as electricity in the year 2000. It is not really accurate to add the contributions of different solar technologies, since they compete with each other as well as with nonsolar technologies, but for now we ignore such competition effects. Taken together, the solar technologies thus display a sort of supply curve that shows that as the cost of competing sources of energy increases, more of the solar options will be adopted, and the total contribution of solar energy will grow. Since the contribution as a function of the cost of competing fuels for each technology is actually a smooth curve, the sum of the contributions will also be a smooth curve.

HEATING AND COOLING

Under this heading we consider the direct collection of solar energy to heat domestic water and heat and air condition the interiors of residential and nonresidential buildings. A discussion of these applications in relation to utility peak loading begins on page 75. A discussion of direct solar heating in commercial processes begins on page 21.

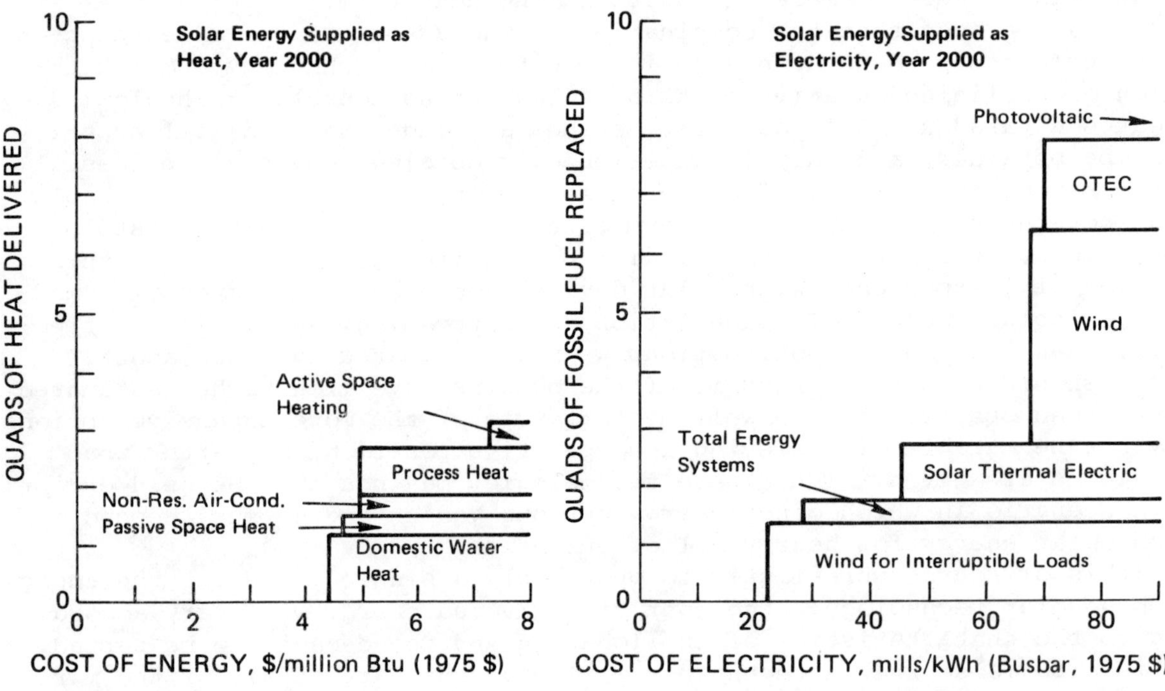

FIGURE IV.1 Estimated energy contributions from various solar energy technologies as a function of the cost of competing fuels for the year 2000.

Of the technologies for direct use of solar energy, domestic water heaters have the greatest potential for quickly gaining a major market. There are several reasons. Hot water is required year-round, thus the investment receives constant use. Solar water heater installations are small compared with space heating installations, so the retrofit task is relatively simple: it is necessary only to make a modest change in the plumbing so that the solar hot water storage tank is placed in the line ahead of the existing conventional water heater, and during the early years of solar market development, when the total number of solar installations is still modest and long-term operating experience is lacking, the fact that the user's investment in a solar water heater is small compared with a solar space heating investment will lead many users to limit their risk by investing only in a water heater.

The potential energy contributions and economics of solar space heating and cooling cannot be understood without considering the important role of energy conservation in buildings. There is little question that given the present and anticipated cost of energy, our building design and construction practices are antiquated. Using currently available materials and advanced design and construction techniques, energy consumption in new homes can be less than half that of typical existing homes with little additional first cost. Throughout our considerations we assume that such energy-conserving practices will be applied to all new homes for which solar energy is considered.

We distinguish between *active* and *passive* solar energy systems for buildings. Passive systems are those in which the transfer of heat from one location to another occurs without the use of mechanical energy. For example, in a typical Florida solar water heater, the transfer of heat from the collector to the elevated water storage tank takes place by natural thermal convection, without the use of a pump. Natural solar heat gain through windows in combination with heat storage in the mass of the building is another example. If movable insulation is used to cover the windows at night, and if the mass of the building is appropriately situated, then one has an effective passive solar heating system. Other passive systems are those in which the collector also serves the functions of heat storage and heat transfer to the building. For example, south-facing concrete walls can be painted black and covered with one or two layers of glass or plastic. The concrete wall transfers heat immediately to the building by natural convection of air past the black surface. The wall also stores excess heat for later use and conducts some heat directly inside. Passive cooling can occur through the use of natural ventilation or through nocturnal radiation cooling (radiation of infrared to the cold night sky).

In contrast to the passive systems, most active solar energy systems require that a surface (roof or wall) be built to keep the sun's energy out of the building, and that a piece of hardware (a solar collector) be mounted on the surface to collect the solar energy and transfer it to a fluid (usually either water or air). A fan or pump moves the fluid through the collector and into a storage container. Later the heat is taken from storage, again by a fan or a pump, for use either to heat the building space directly, or as a source of low-grade heat for a heat pump.

Active systems generally provide much better control of the use of solar heat in a building but at a substantial increase in complexity and first cost.

For a variety of reasons, passive systems have not received the same attention in government programs and in the private sector that has been devoted to active systems. However, it is generally agreed in the solar community that passive methods can be simpler and less costly per unit of energy saved and have a potential for providing acceptable levels of comfort control. Most solar proponents, therefore, advocate proper thermal design of the building first, through the use of energy conservation measures, then design for passive uses of solar energy, and finally application of active solar systems. We assume, in our estimates of energy contributions and costs, that these measures gain a market in this order as the cost of fuels and electricity increases.

The energy contributions and the economics of solar heating and cooling depend strongly upon the local climate's heating and cooling loads and available insolation. We do not provide detailed descriptions of these regional variations, since such detail cannot be justified for the purpose of this report. Instead, we base our estimates upon a typical house in a typical climate. It should be understood that the results we obtain are indicative of correct values, but are not necessarily the values that will be obtained even as averages in any specific scenario of application of solar heating and cooling.

The example house has an area of 1,500 square feet (three bedrooms) and is located in a 5,000 degree-day heating climate. (Washington, D.C., has about 4,500 heating degree-days.) This house, constructed to the old FHA thermal performance standards, would use about 120 million Btu per year for space heating. Through relatively simple and inexpensive energy conservation methods, the load can be reduced to about 60 million Btu. We assume that such energy-conservation measures are always included in the design before solar energy is considered for use on a building.

This house would, in conventional design, have 50 square feet of south-facing glass. In adding passive solar energy features, this is increased to 100 square feet (with storm windows or thermopane, and with sun shades for the summer) to provide increased direct heat gain from the winter sun into the living space. All glass is provided with interior movable insulating panels for use at night. This is a minimal passive system. Much larger energy savings are certainly possible by more sophisticated measures. The range of such possibilities of passive design is only now being explored. Studies now under way to collect performance data on existing successful passive designs, and to analyze how these and other possible designs would perform in a variety of climates, will allow a much better estimate to be made a year hence of the potential impact on a national scale of passive solar systems.

Through energy-conserving design and the minimum passive solar design specified above, the active solar energy system to provide 68 percent of remaining heating load for this house requires only about 300 square feet of collector area. Even less area might be required if the active solar system uses a heat pump. Because of this small size, the problems of including the collectors in the building design are simplified and installation cost per square foot is reduced from what it would be in a house built to old FHA standards.

Table IV.1 summarizes some typical extra initial (1975) costs and energy contributions of this sequence of measures.

The cost of this active solar energy system corresponds to present costs for what are essentially prototype systems. This cost of $20 per square foot of collector area is expected to come down as contractors gain more experience with such systems. We assume that by the year 1990 the cost will be down to $10 per square foot and will remain constant thereafter.

The assumption that the house will use a proper energy-conserving design for its local climate reduces the possible contribution by an active solar cooling system. In homes of present construction with little or no insulation, central air conditioners consume 5,000 kWh or more of electricity per year. Proposed efficiency standards would require that these units have an Energy Efficiency Ratio (EER) of 10, instead of the present average of 8.5. Increasing insulation of the home and reducing infiltration cuts the heat gain to the building. We assume that as a result of better insulation, shading, and venting of heat loads to the outside of the building, new housing will require on the average only 1,500 kWh per year for central air conditioning. Of course, this number is extremely dependent upon local conditions. But on the average, the assumption of low cooling load reduces the potential energy contribution of an active solar cooling system and thus worsens its economics. However, there is also a significant possibility for effective designs for low-cost passive cooling that should be the goal of new R&D efforts.

Air conditioning for both residential and nonresidential application will be the slowest developing solar application. (We no not include here solar heat pump systems, since in their air conditioning mode they make no use of solar energy.) Residential air conditioning applications will grow more slowly than nonresidential. This presents a contrast to the water heating and space heating market, in which residential applications are expected to grow the most rapidly. The cooling season for

TABLE IV.1 Potential Costs and Contributions of Conservation Measures in Residential Solar Construction

Action	Extra Initial Cost (1975 $)	Annual Energy Contribution (million Btu)	$ per million Btu Contributed per Year (1975 $)
Energy-Conserving Design	3,000	60	50
Passive Solar Design	450	7.5	60
Active Solar Energy System	6,000	36	167

commercial buildings such as stores and office buildings is longer than the residential cooling season. Thus, in these important markets, the solar equipment will be better utilized than in residential applications. In the southern part of the country, air conditioning of stores and office buildings is very nearly a year-round load. The majority of nonresidential air conditioning systems are chilled water systems in which the "product" of the air conditioning machinery is chilled water. These systems lend themselves to retrofit with a solar unit in which the solar collectors and heat-driven chiller supplement the existing conventional equipment. In some cases, only the solar collection system must be added since a substantial number of heat-actuated chillers are now in use. Finally, the typical solar-driven, heat-actuated chiller requires the use of a wet cooling tower. Wet cooling towers are now widely used for heat rejection in commercial air conditioning installations but are not generally used in residential installations. The operating complexity, maintenance requirements, and water consumption imposed by the wet cooling tower are not material deterrents to the use of solar air conditioning in nonresidential applications, but are meaningful deterrents in the case of residential applications, particularly in single-family residences.

There is no question that solar heating and cooling systems can be built and operated in an efficient manner. However, the economics of solar heating and cooling remain somewhat uncertain. In general, six major factors influence the economic viability of solar systems: (1) The initial installed cost of the solar system, (2) available solar flux at the site, (3) energy utilization by the system load, (4) local prices of alternative fuels, (5) cost of backup power for a solar system, and (6) interest rate and taxes charged the owner of the system.

The last of these factors is of great significance, and provides opportunity for federal and state actions to advance use of solar heating and cooling. The Solar Energy Industries Association has proposed a tax credit to homeowners equal to 40 percent of the first $2,000 invested and 25 percent of the next $6,000 invested in solar equipment. The reason for proposing a larger credit for the first $2,000 investment is twofold. In the first place, it will provide the largest incentive to the owner of the relatively small home who needs a relatively small and inexpensive installation. The owner of the smaller home is characteristically in a lower income bracket than the owner of the larger home and needs help more. Secondly, a 40 percent credit on the first $2,000 will particularly provide an incentive to the installation of solar water heaters. These are the most productive in terms of energy savings. They also will lead to general acceptance of solar energy, and in turn their broad acceptance will help with subsequent acceptance of solar energy for space heating.

We believe that the tax treatment proposed can be justified. The homeowner who purchases and installs residential solar equipment becomes an energy producer similar to the utility company which purchases and installs new generating capacity, the coal mining company which purchases and installs mining equipment, or the oil and gas company which drills an oil or gas well or builds a refinery or pipeline. All of these industries receive tax credits for investment, in common with other industries.

Indeed, the only exception is the homeowner, who with a solar system becomes an energy producer, but receives no special tax treatment under existing legislation. We therefore suggest an equivalent tax credit for the energy producing homeowner.

It should also be expected that some consumers will choose to purchase a solar energy system for a new building even if it does not seem justified by the prevailing costs of fuel at the time. This is because the system, once purchased, is a form of insurance against rapidly growing fuel costs. The individual who builds a new home without including a solar system faces the prospect of either continuing to pay whatever price might be charged for fuels during the life of the building, or else paying the much higher cost of retrofitting a solar system to the building at a later time.

Table IV.2 reports our estimates of the energy contributions and costs of solar heating and cooling. The cost estimates do not include the effects of the tax policy suggested above.

PROCESS HEAT

The particular importance of using solar energy for process heat has not been widely understood, and this application has unfortunately lagged far behind heating and cooling of buildings in government programs. However, this is now a critical opportunity for solar energy because about 50 percent of industrial and agricultural process heat presently comes from natural gas, the price of which is expected to rise sharply over the next few years. Perhaps more important, industrial and agricultural users of gas are being put on interruptible service or are being terminated, and it is impossible in many areas for a new plant to obtain gas service. Since plants must have a dependable and continuous fuel supply, a shift from gas to oil is now occurring, and this shift is expected to accelerate over the next few years. But as oil prices continue to rise, industry will no doubt begin shifting within the next 10 to 20 years to using coal for process heat. Because the use of coal is more labor intensive and requires handling and storage facilities, the shift to coal will be easier for large plants (those with annual process heat requirements greater than about 1 trillion Btu) than for small ones.

Two requirements must be met before industry will adopt solar energy on any substantial scale as a supplemental source of process heat. First, the technical feasibility of solar heat must be demonstrated. Will the collectors provide the predicted amount of heat reliably over their expected lifetime? Will operation and maintenance costs be reasonable? How much storage is required? Can the solar collectors and storage system be wedded to an existing fossil fuel facility so that there will always be a reliable backup source of process heat?

The second and equally important requirement is economic feasibility. The collectors and storage system will entail a large capital investment. A company will require a minimum rate of return on that investment, normally in the range of 8 to 12 percent, excluding an inflation factor.

A life-cycle cost analysis of the hybrid solar/fossil fuel system must take into account (1) annual cash savings in fossil fuel; (2) depreciation

TABLE IV.2 Potential Energy Contributions and Costs, Solar Heating and Cooling

System Type	1975	1985	1990	2000	2010
Water Heating[a]					
Energy Contribution, Quads[b]	--	0.1	0.4	1.2	1.3
Number of Systems, Millions	--	12	29	87	92
System Cost, $[c]	1,200	900	700	700	700
Energy Cost, $/Million Btu[d,e]	7.50	5.70	4.40	4.40	4.40
Space Heating					
Passive Systems[f]					
Energy Contribution, Quads[b]	--	0.1	0.2	0.3	0.4
Number of Systems, Millions	--	15	22	37	52
System Cost, $[c]	450	400	400	400	400
Energy Cost, $/Million Btu[d]	5.20	4.60	4.60	4.60	4.60
Active Systems[f]					
Energy Contribution, Quads[b]	--	0.02	0.06	0.37	0.95
Number of Systems, Millions	--	0.7	1.8	10	26
System Cost, $[c]	6,000	3,600	3,000	3,000	3,000
Energy Cost, $/Million Btu[d]	14.50	8.70	7.30	7.30	7.30
Solar Air Conditioning					
Residential[g]					
Energy Contribution, Quads[h,j]	--	0.006	0.007	0.05	0.12
Number of Systems, Millions	--	0.06	0.7	5	12
System Cost $[c,i]	2,500	1,500	1,000	700	700
Energy Cost, Mills/kWh[e,j]	110	130	87	61	61
Nonresidential					
Energy Contribution, Quads[k]	--	0.003	0.06	0.36	1.50
Total Cooling Capacity, Million Tons[l]	--	0.04	0.8	5.1	22
System Cost Per Ton, $	4,200	3,100	2,500	2,500	2,500
Energy Cost, $/Million Btu[m]	8.00	6.00	5.00	5.00	5.00

[a] Includes only provision of domestic hot water. Commercial hot water is included under industrial process heat.

[b] Quads actually delivered as useful heat. Each such Btu is equivalent to approximately 1.7 Btu of fossil fuel burned at normal (60 percent) efficiency.

[c] Declining costs in 1975 dollars are assumed to be due to a learning-curve effect. If sales are significantly less than indicated the reduction in cost may not be realized as soon. This is cost in addition to cost for conventional backup system.

[d] One million Btu from the solar system is equivalent to 1.7 million Btu of fossil fuel, so at $4.40 per million Btu solar is competitive with fossil fuel at $2.65 per million Btu.

TABLE IV.2 Continued

---

[e]The real cost of capital (no inflation) for the homeowner is taken to be 6 percent with a 20-year mortgage. No tax effects are included, so the annual payment is 8.72 percent of the first cost.

[f]It is assumed that new energy-conserving homes will require on the average 60 million Btu per year of space heat so the active solar system needs only 300 square feet of collector area.

[g]It is assumed that new energy-conserving homes will require on the average 1,500 kWh per year to drive a vapor compression air conditioner with an EER of 10.0.

[h]Converted from kWh at 10,000 Btu per kWh.

[i]This is the additional cost over cost of a vapor compression air conditioner and over cost of an active solar space heating system. The cost in the year 2000 is based on the assumption that in mass production a Rankine-driven air conditioner will cost no more than double the present cost of a vapor compression air conditioner (about $700).

[j]Assumed contribution of the solar system is to displace use of 1,000 kWh per year.

[k]Energy content of fossil fuel displaced by solar heat input to absorption chillers.

[l]Assuming an average coefficient of performance of 0.5.

[m]Based on 12 percent real cost of capital and $50 per ton annual operating and maintenance cost.

of solar system; (3) increase in income tax based on net additional income from fuel savings; (4) initial investment tax credit; (5) increased operation and maintenance costs due to solar system; (6) expected replacement time for certain collector components, e.g., plastic water bags; (7) added pumping costs to circulate the solar-heated water, air, or steam through the collector field; and (8) possible salvage value of solar facility. The resulting energy cost, which applies to the overall solar/fossil fuel system, can then be compared with the energy cost for the fossil fuel system operating alone. If both technical and economic feasibility can be demonstrated, then it can be expected that many companies will lessen their dependence on fossil fuels in the coming years and consider other alternatives such as installing supplementary solar heating systems. Such systems will have the advantage of reducing the logistic problems and uncertainty in maintaining adequate oil or coal supplies.

Because of the importance in industry of maintaining a favorable cost position vis-a-vis the competition, it is unlikely that solar systems will be acceptable if they result in an increase in annual energy costs. Motives of fuel conservation or reduction of environmental effects cannot be relied upon to play a significant role. However, there is an important subjective factor that may influence a company's decision to use solar energy. If it is expected that the price of the particular fossil fuel in current use will rise at a rapid rate or that there are uncertainties about a continued reliable supply of that fuel, then a company might decide to install a solar system to lessen its dependence on fossil fuel, even if it would clearly result in an immediate increase in energy costs. This is a subjective element that will vary in coming years and from one company to another. Although it is an element in an alternative scenario discussed later, it is ignored in our baseline scenario.

The Industrial Requirements for Process Heat

The percentage of total U.S. energy used by industry has fallen over the past 25 years--from 44 percent in 1950 to 40 percent in 1975 (Dow Chemical, 1975). We assume that this will remain constant at 40 percent over the next 35 years. There have been many projections of total U.S. energy consumption over the next 25 to 50 years. Since the beginning of the energy crisis and the emphasis on conservation, these projections have become more conservative. A middle-of-the-road projection is used here, namely that overall energy use will increase at a 2.7 percent annual rate over the next 35 years (Center for Strategic and International Studies, 1973). This corresponds to a nearly twofold increase in energy use by the year 2000.

The first column of Table IV.3 is a breakdown of total process heat use in 1975 among the six leading industrial users. These six industries use almost 80 percent of all process heat. The remainder is used by many different types of plants, generally of smaller size.

TABLE IV.3 Process Heat Use, in Quads, Among Six Leading Industrial Users in 1975

Type of Industry	Total Use[a]	Percentage in Steam/ Hot Water	Energy in Steam/Hot Water	Best Solar Candidates
Primary Metals	4.2	21	0.9	
Chemicals	4.0	60	2.4	2.4
Petroleum	1.9	35	0.7	
Stone, Clay, and Glass	1.7	8	0.14	
Paper and Pulp	1.7	85	1.4	1.4
Food Processing	1.3	~90	1.2	1.2
All Others	4.2	~73	3.1	3.1
Grand Totals	19.0		9.8	8.1

[a] Dow Chemical (1975); Unger (1975).

Three conversion systems account for most of industrial fuel use: generation of steam or hot water, open kiln firing, and process heaters and furnaces. Solar energy can be expected to be most important in the first category. Therefore, the second column of Table IV.3 lists the percent of each industry's process heat use that is in the form of steam or hot water and the third column lists the absolute amounts. The primary metals, stone, clay, glass, and petroleum industries are poor solar candidates because of the high temperatures required in the first two industries and the heavy use of process residual energy in the petroleum industry. Therefore, we will exclude these three industries from further consideration. The last column of Table IV.3 lists the steam or hot water process energy used by the remaining industries.

The best solar candidates constitute 43 percent of all process heat use. However, we should also include hot air, particularly for drying applications in agriculture and the food-processing industry. Although we do not have firm figures for these latter applications, we assume an approximate value for them that results in a total of 50 percent of all process heat used in a form favorable to supply by solar energy systems, or 12.5 percent of total U.S. energy usage.

Because the capital costs of solar systems are high, we assume that for quite a few years--in fact 35 years for present purposes--such systems will be adopted in only the sunnier regions of the country. The annual solar radiation in the western states is from 50 percent to 100 percent greater than in the Great Lakes or North Atlantic states. An average of about 40 percent of all process heat is used in the South Central, Mountain, and Pacific states (Dow Chemical, 1975). This share can be expected to increase over the next 35 years because of the continued shift of population and industry to the South and West. Therefore we assume that an average of 50 percent of the U.S. process heat requirement will occur in "sunshine" regions. It will also be assumed that the percentage of process heat in the form of steam, hot water, or hot air is the same in all regions of the country. This would seem to be a conservative assumption since the primary metals and stone, clay, and glass industries, which are not viable candidates for solar systems, are concentrated in the eastern half of the country.

Table IV.4 sums up the U.S. total energy use and industrial process heat use over the next 35 years as projected by this scenario. Because of the lack of more complete data on type and geographical distribution of process heat use, these projections are probably the best that can be made at this time. However, the government has sponsored studies of industrial process heat and its distribution by industry, region, temperature, etc. These studies will allow better estimates to be made in the future.

It is to be noted that in Tables IV.3 and IV.4 there is no explicit mention of agricultural process heat. The largest energy use in this category is drying grains and other crops, such as onions, prunes, raisins, etc., as well as dehydration of dairy products. Most of these processes take place outside the farm gate and hence are included under the category of food processing.

We have not considered here the potential for energy conservation in industry. Effective conservation actions, especially recovery of waste

TABLE IV.4 Projected U.S. Total and Process Energy Use, in Quads, Assuming a 2.7 Percent Per Year Growth Rate in All Categories Over Next 35 Years

Use	1975	1985	1990	2000	2010
1. Total Energy Use	75	98	112	146	190
2. Total Industrial Use (40 Percent of Line 1)	30	39	45	58	76
3. Total Process Heat (25 Percent of Line 1)	19	25	28	36	48
4. Steam/Hot Water/Hot Air for Solar-Favorable Industries (50 Percent of Line 3)	9.5	12.5	14	18	24
5. Steam/Hot Water/Hot Air for Solar-Favorable Industries in "Sunshine" Areas (50 Percent of Line 4)	4.7	6.2	7	9	12

heat, should always be instituted before considering the use of solar process heat. Because of its relatively high costs, solar energy should be used only in energy-efficient systems.

Some Characteristics of Industrial Process Heat Solar Systems

Solar collectors to heat water or air to low temperatures (<210°F) are simple, well-understood flat-plate collectors. An industrial solar process heat project now under way in New Mexico will, it is hoped, demonstrate both the reliability and cost effectiveness of large-area, low-temperature solar water heaters (Lawrence Livermore, 1976).

Using solar energy to make process steam (say at a typical required temperature range of 300-350°F) is a more difficult task. The solar steam collectors must be combined with a fossil fuel backup system to provide steam on demand. When combustion of fossil fuel is used, one Btu of steam costs no more to produce than one Btu of hot water. The range of annual average collection efficiencies for a low-concentration collector producing steam can be expected to be comparable to that of a flat-plate collector for water heating (40-50 percent). It follows that the allowed system cost per square foot of collector must be about the same for the two in order for them both to be competitive with fossil fuel.

Since both solar process steam and hot water systems must generally be provided with 100 percent fossil fuel backup capability, they are supplementary systems to be thought of strictly as fuel savers. In many cases it will be most cost-effective to provide minimal or even zero storage capability to reduce system costs.

According to the Dow Chemical study, 90 percent of all industrial steam and hot water systems produce steam at 200 pounds per square inch gauge (~390°F) or above (Dow Chemical, 1975). Other estimates indicate that perhaps one-third of "steam" process heat is in the form of hot water. If a plant requires steam, a solar system could be used to preheat the water before sending it to the fossil fuel boilers. However, in this case the maximum contribution of the solar system would be only about 15 percent of the total required process energy.

Many plants require only hot water but find it more convenient to produce steam, pipe it to the final use point, and convert it to hot water. In other cases the plant requires steam at various temperatures as well as hot water. It is then most convenient to produce all the steam at the highest required temperature. It may be concluded that if solar energy systems are to make a major contribution to providing process heat, cost-effective collectors to produce steam, as well as collectors to produce hot water and hot air, must be developed.

Table IV.5 shows a sample calculation of annual costs for low-temperature process heat in a conceptual design developed for an ore-milling plant in the Southwest that uses 200 billion Btu of hot water per

TABLE IV.5 Comparison of Annual Costs, Low-Temperature Water Heating

Plan A (Fuel Oil Only)		Plan B (50% Solar/50% Fuel Oil)	
Cost of Oil	= $738,920	Cost of Oil	= $369,460
Boilers $60,000 × (Capital Recovery Factor: 8%, 20 Yrs.)	= 6,110	Solar Facility $2,090,880 × (Capital Recovery Factor: 8%, 20 Yrs.)	= 212,960
Operation and Maintenance	= 20,000	Boilers	= 6,110
	$765,030	Operation and Maintenance	= 40,000
		Extra Pump Power	= 2,000
		Added Income Tax	= 122,460
			$752,990
Cost of Energy Per Million Btu	= $3.83	Cost of Energy Per Million Btu	= $3.76

Conditions:

1. Annual requirement for process hot water = 200 billion Btu.
2. Annual contribution from solar facility = 100 billion Btu (50 percent).
3. Price of fuel oil = $15 per barrel (assume 70 percent conversion efficiency).
4. Area of solar collectors ≃ 8 acres.
5. System cost for solar facility ≃ $6.00 per square foot. ($6.00 per square foot × 348,480 square feet = $2,090,800.)
6. Required rate of return on solar investment = 8 percent.
7. Straight-line depreciation over 20 years.
8. No salvage value.

year. If fuel oil at an onsite delivered cost of $15 per barrel is used, the unit cost of energy comes to $3.83 per million Btu. If a combined 50 percent solar/50 percent fuel oil system is used under the set of listed conditions, the overall cost of energy is $3.76 per million Btu. Hence, at the assumed solar system cost of $6 per square foot, there results a slight decrease in the annual cost of energy.

In this example, a real rate of return is used which does not include any general inflation factor. Prices, costs, and wages are all assumed to inflate at the same rate. However, if it is assumed that fossil fuel prices will rise at a rate differentially higher than the inflation rate, then the allowed solar system cost will rise.

A plant using 200 billion Btu of process heat per year is classified in the lower third of process heat users. This would correspond to 35,000 gallons per hour of 140°F water or 20,000 pounds per hour of 350°F steam. Yet the investment of $2 million in a solar system is a substantial one. In addition, about 15 acres of real estate or factory rooftop must be available for the eight acres of collectors. (The cost of the real estate has been ignored in the calculation since land at $2,000 per acre is equivalent to only 4¢ per square foot.) A medium-size industrial process heat user might require 100,000 pounds of steam per hour representing a $10 million investment even if concentrating collectors are so inexpensive that a $6 per square foot solar system can supply half that steam. About 75 acres of land or rooftop would be required.

Input Assumptions for Baseline Scenario

The baseline scenario for projecting the use of solar energy for process heat over the next 35 years is based on the following set of assumptions:

(1) Solar hot water systems will be available in a year or two at costs of $5 to $8 per square foot (total system costs per square foot of collector in 1976 dollars). It is assumed that economies of scale allow such systems to be less expensive per square foot than active residential space heating systems. Solar low-pressure steam systems will be available in this cost range by 1985. It might be argued that these conditions are too stringent because of the probability that oil prices will rise more rapidly than the inflation rate. However, it must be noted that capital costs have historically been rising at a rate of 2.8 percent faster than inflation (Dow Chemical, 1975). Also, if oil prices become unacceptably high and solar systems remain too expensive, plants will tend to choose the option of converting to coal.

(2) Between now and 1985 the average solar process hot water or hot air system will produce 100 billion Btu per hour. At a system cost of $10 per square foot, this represents a multimillion dollar investment. By 1985, 1,000 systems of this size will be operating in the "sunshine" regions. The total investment in these systems will be more than $2 billion and the annual energy contribution about 100 trillion Btu (0.1 quad).

(3) Beginning in 1985, solar systems for low-pressure steam will become cost effective. For all types of solar process heat systems for the next 25 years, a lower limit growth rate of 10 percent per year and an upper limit growth rate of 15 percent is assumed.

TABLE IV.6 Solar Process Heat Contributions, in Quads, Over Next 35 Years According to Baseline Scenario and Two Alternative Scenarios

Category	1975	1985	1990	2000	2010
Total U.S. Energy (2.7 Percent Growth Rate)	75	98	112	146	190
Steam/Hot Water/Hot Air for Solar-Favorable Industries in "Sunshine" Areas	4.7	6.2	7.0	9.0	12
Solar Contribution					
Baseline Scenario (10 Percent Growth Rate)	0	0.1	0.16	0.42	1.1
Baseline Scenario (15 Percent Growth Rate)	0	0.1	0.20	0.81	3.3
Pessimistic Scenario	0	0.02	0.05	0.1	0.2
Accelerated Scenario	0	0.2	0.4	1.6	6.6

Table IV.6 repeats some of the data given in Table IV.4 and gives the projected solar energy contributions for the lower and upper limits of the baseline scenario. At the 10 percent growth rate the total value of installed solar process heat systems in 2010 would exceed $20 billion. At the 15 percent growth rate it would exceed $60 billion.

These estimates are shown in Figure IV.2. The lower limit of the baseline scenario band predicts that in 2010 about 20 percent of the process heat used in solar-favorable industries situated in "sunshine" areas would be contributed by solar energy. The upper limit estimate is nearly 30 percent contribution by solar energy. A middle-range guess would be two quads contributed by solar energy in 2010. This would correspond to about 1 percent of the projected total U.S. energy use in 2010 or about the equivalent of 340 million barrels of oil per year. It would also represent a total investment in solar process heat systems of $40 billion to $60 billion (1975 dollars) over the next 35 years.

Alternative Scenarios

Two alternative scenarios can well be envisioned for the future of solar process heat systems.

Accelerated Scenario

If it is assumed that, because of increased world demand or international political complications, our access to oil becomes severely limited over the next 5 to 10 years, then industry would turn to coal sooner than planned. (Just as industry has been given a lower priority for natural

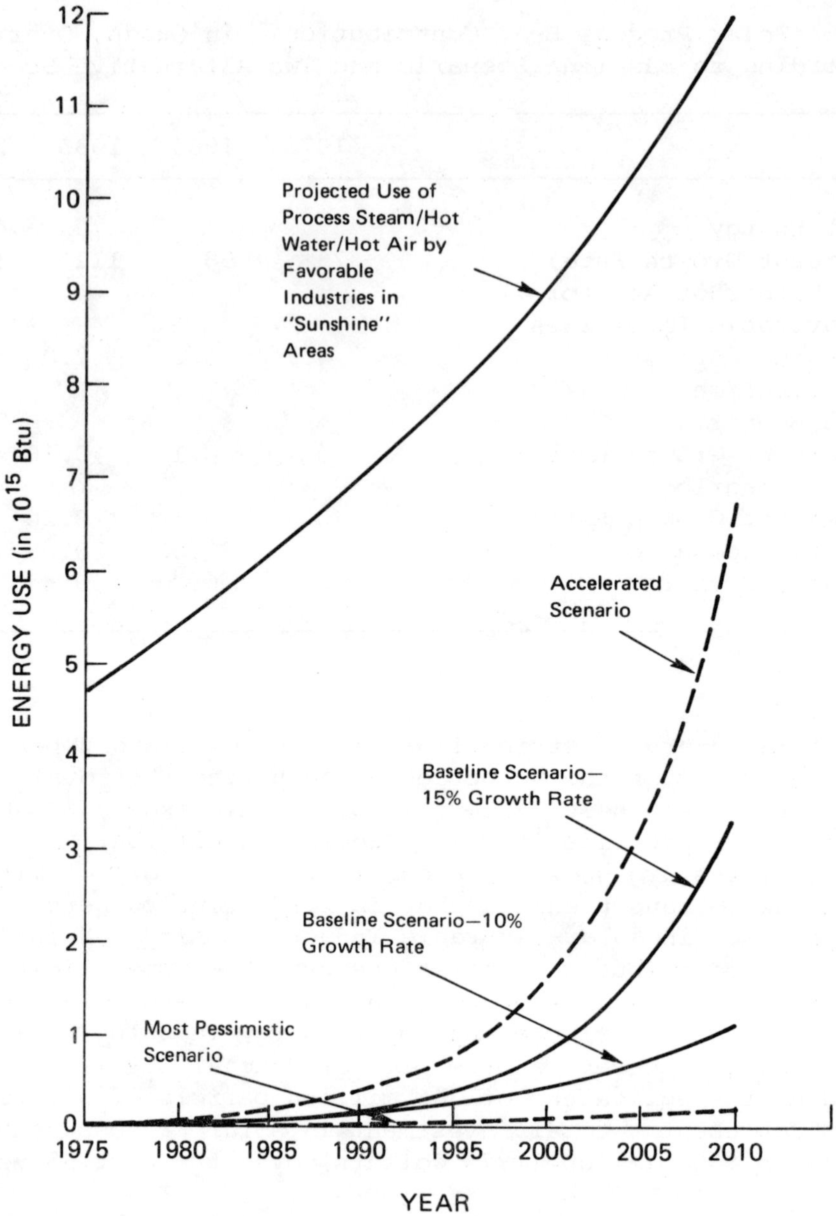

FIGURE IV.2 Scenarios for the energy contributions by solar industrial process heat.

gas than has residential heating, it also would probably be given a lower priority for scarce oil than would the private automobile.) In addition, if it is assumed that critical problems arise in obtaining and using western coal (e.g., unavailability of transportation for great quantities of coal, objection to massive strip-mining, labor shortages, or various environmental restraints), then industry would by necessity turn to solar systems to reduce their dependence on oil or coal. Under these conditions we have assumed twice the contribution of solar systems as predicted by the favored scenario with a 15 percent growth rate. Hence, in 2010 about half (6.6 quads) of the potentially favorable market for solar process heat systems (12 quads) would be supplied by such systems. (See Table IV.7.)

Most Pessimistic Scenario

If the initial generation of demonstration solar process heat systems proves too expensive or has poor reliability, industry may give up on

TABLE IV.7 Potential Energy Contributions and Costs, Process Heat

Process Heat	1975	1985	1990	2000	2010
Energy Contribution[a]					
Baseline Scenario, Quads[b]	--	0.1	0.2	0.8	3.3
Accelerated Scenario, Quads[c]	--	0.2	0.4	1.6	6.6
Total U.S. Process Heat, Quads[d]	(19)	(25)	(28)	(36)	(48)
Suitable for Solar, Quads[e]	(4.7)	(6.2)	(7)	(9)	(12)
Cost of System, $ Per Square Foot[f]	6.00	8.00	8.00	8.00	8.00
Cost of Energy, $ Per Million Btu[g,h]	3.76	5.00	5.00	5.00	5.00

[a] There is no break-out of agricultural process heat because most of such heat is used for drying of grains and various crops, and is included in the food-processing industry.

[b] Energy content of fossil fuel displaced by input of solar heat.

[c] This is simply double the baseline scenario.

[d] This is the assumed total U.S. requirements for both solar and nonsolar process heat used at any temperature.

[e] This is the assumed total amount of process heat used as steam, hot water, or hot air by industries with processes adaptable to solar and located in "sunshine" areas.

[f] The 1975 cost is the approximate present cost of a system to provide process hot water in a favorable application. The costs for 1985 and later are estimated costs for systems for low-pressure steam.

[g] It is assumed that until 1985 only hot water and hot air will be available at this cost, but after that time development of low-cost concentrating collectors will make steam available at only slightly higher cost.

[h] Real rate of return of solar investment (no inflation) is assumed to be 8 percent with straight-line depreciation over 20 years.

solar energy as an alternative energy source and make long-range plans for the use of oil and eventually of coal. If, in addition, oil prices stabilize over the next few years and "cheap" western coal becomes readily available, there will be little incentive to invest in solar systems. Some solar systems will still be built in very favorable locations and where the clean heat furnished by solar energy is attractive. Here we have assumed that by 2010 only about 1.5 percent of the potentially favorable process heat market would be supplied by solar systems. Even this would represent an annual saving of 34 million barrels of oil equivalent.

Needed Research and Development

For the predictions of the baseline scenario to become reality, top priority should be given to developing a family of large area collectors specifically for process hot water, hot air, and low-pressure steam. Installed system costs must be held to about $8 per square foot of collector. Actual collector costs should be held to about $4 to $6 per square foot. Present collector costs are too high for the scenarios presented here to be met.

The experience to date with flat-plate and low-concentration collectors fabricated from metal and glass provides little hope that such collectors can be built at these low cost levels. The only alternative is to use lightweight, strong, low-cost polymeric materials (for covers on flat-plate collectors, and for reflectors on concentrating collectors) that have lifetimes of five years or longer when exposed to sun and weather. There are a number of potentially good materials of this type already on the market but there remains a need for considerable improvement.

There should be a major effort to develop low-cost collectors for producing 300-350° steam. Very little work has been done in this area up until now. The long-term impact of solar energy in providing process heat will depend to a large extent on the availability of very low-cost solar collectors to produce steam. Also there should be a coordinated effort made with a number of "solar-favorable" plants representing different industries to design solar steam systems that can be combined with already existing fossil fuel systems. Determination of minimum storage requirements and development of appropriate storage systems will be essential parts of this effort.

BIOCONVERSION

Biomass conversion, or bioconversion, was the major--often the exclusive--source of energy for man until a century ago. In 1875 wood still provided two-thirds of the total energy consumed in the U.S. Bioconversion offers many advantages over other techniques for the entrapment of solar energy but it suffers from the sequential nature of the overall process and the resultant multiplication of efficiencies. The essential steps in any bioconversion process and the coupling mechanisms between them are summarized in Figure IV.3.

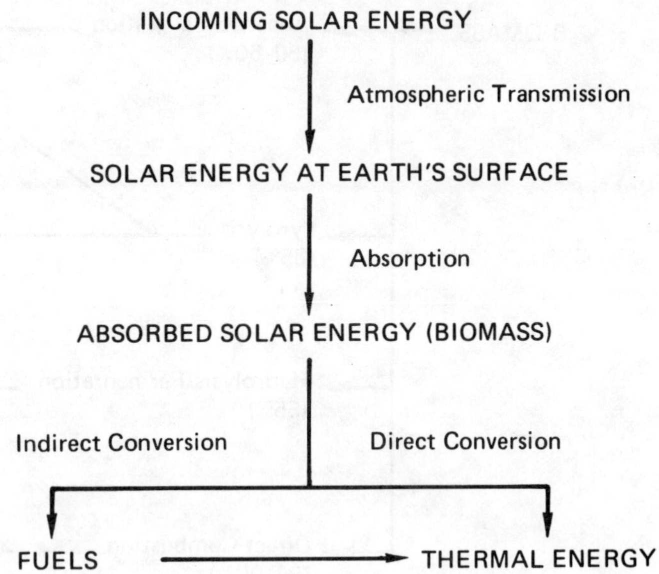

FIGURE IV.3 Essential steps in the bioconversion process.

Absorbed solar energy is present in the tissues of plants and certain microbial forms (biomass) as chemical (bond) energy. Carbohydrates, fats, and proteins are the primary reservoirs of stored solar energy, but other types of compounds, e.g., hydrocarbons, are secreted abundantly in some species.

The energy available in plant tissue and other biomass forms can be made available either directly, through combustion of the tissue, or indirectly. Indirect conversion methods which have been proposed and studied are:

(1) *pyrolysis* to mixed liquid (oils) and gaseous fuels,

(2) *anaerobic fermentation* to gaseous fuels (methane, hydrogen, carbon monoxide), and

(3) *hydrolysis* (of carbohydrates) to glucose followed by fermentation of the glucose to ethanol.

A summary of these procedures, with the projected *maximum* energy recovery factor for each, is given in Figure IV.4.

Indirect conversion of biomass, either to liquid or gaseous fuels, adds another step--and one more recovery factor--to the chain. On the other hand, these routes offer fuels which are adaptable to certain energy requirements that cannot be served by direct combustion of biomass, e.g., internal combustion engines.

Existing and Potential Sources

Four significant sources of biomass for conversion have been proposed:

(1) municipal solid wastes,

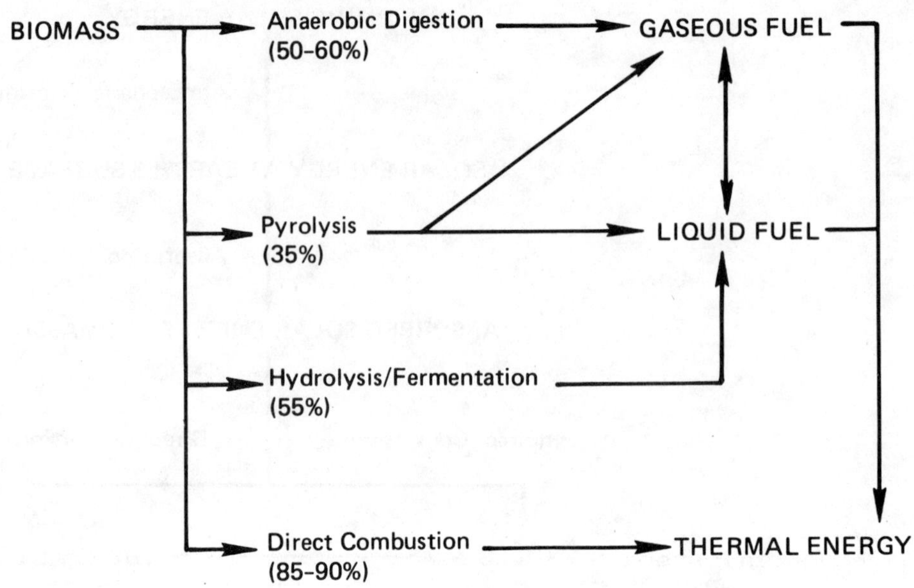

FIGURE IV.4 Summary of indirect and direct bioconversion methods.

(2) agricultural residues,
(3) terrestrial energy farms, and
(4) marine energy farms.

Potential future energy contributions from these four sources are summarized in Table IV.8. Each projection is presented as the product of total energy content available and a recovery factor. This division has been introduced because, in many cases, the two components will respond differently to technological development and socioeconomic changes. For example, a change in agricultural practice from extreme dependence on petroleum-based fertilizers to greater use of residue recycling will reduce the availability of residues even though agricultural production continues to increase. Specific comments on the four biomass categories follow. These comments relate to:

- nature and availability of required technology
- prospective changes in the resource base
- reliability of estimates
- potential problems

Municipal Wastes

Technology for the recovery of energy from municipal wastes is already established. Improvements in operating efficiencies may be expected but these will be small. The projections in Table IV.8 reflect a belief that socioeconomic pressures will result in a gradual reduction in the

TABLE IV.8 Potential Energy Contributions and Costs, Bioconversion

Biomass Source	1975	1985	1990	2000	2010
Municipal Wastes					
Energy Content, Quads[a]	(1.0)	(1.6)	(2.0)	(2.2)	(2.0)
Recovery Factor[b]	0.95	0.95	0.95	0.95	0.95
Available Energy, Quads	0.95	1.5	1.9	2.1	1.9
Cost of Energy, $ Per Million Btu[c]	2.10	2.10	2.10	2.10	2.10
Agricultural Residues[d]					
Energy Content, Quads[a]	(8.0)	(9.0)	(9.5)	(10.0)	(10.0)
Recovery Factor[e]	--	0.25	0.30	0.35	0.35
Available Energy, Quads	--	2.2	2.8	3.5	3.5
Cost of Energy, $ Per Million Btu[f]	2.00	2.00	2.00	2.00	2.00
Energy Farming, Terrestrial					
Energy Content, Quads[a]	--	0.5	1.5	3.5	4.0
Recovery Factor[g]	--	0.75	0.80	0.85	0.85
Available Energy, Quads	--	0.4	1.2	1.9	3.4
Cost of Energy, $ Per Million Btu[h]	1.00	0.55	0.40	0.30	0.25
Cost to Compete with Food, $ Per Million Btu[i]	2.90	?	?	?	?
Energy Farming, Marine					
Energy Content, Quads[a]	--	0.015	0.03	0.65	1.4
Recovery Factor[j]	--	0.7	0.75	0.75	0.75
Available Energy, Quads	--	0.01	0.02	0.5	1.1
Cost of Energy, $ Per Million Btu[k]	(3.00 to 9.00)				

[a] Energy content of raw biomass to be used directly as a fuel.

[b] Efficiency of recovery of raw biomass energy through direct combustion.

[c] Based on actual values for currently operating plants; projections assume no significant changes in technology.

[d] Agricultural residues include crop residues not consumed in crop production (recycled); forest residues; and manure.

[e] Recovery factor includes partial recovery of available wastes and conversion to fuel gas (600 Btu per standard cubic foot) by anaerobic digestion.

[f] Initial assumptions are 10 percent return on investment, 25 year payback, and 10 year equipment depreciation. Projections assume an increase in conversion efficiency and an increase in the required return on investment to 15 percent by 2010.

[g] Recovery factor includes only direct combustion of raw biomass.

[h] Initial assumptions are yield of 5 tons per acre of recoverable biomass, land cost of $50 per acre, interest and taxes equal to 10 percent of the land cost, and harvest cost of $9 per ton. Projections assume and

TABLE IV.8 Continued

---

increase in yield to 30 tons/acre by substitution of superior crops and a parallel increase in nutrient costs and in interest and taxes to 15 percent by 2010.

[i] The equivalent cost of energy that corresponds to a return on use of the land, water, and nutrients equal to that from production of food. Based on average U.S. yield of corn (72 bushels per acre) and average price for corn ($3.20 per bushel) in 1974.

[j] Recovery factor includes conversion to fuel gas (600 Btu per standard cubic foot) by anerobic digestion.

[k] Costs of Wilcox and Leese (1976). Proposal is entirely hypothetical so no projection of variation of costs with times has been attempted.

volume of such wastes by the end of this century, but such a turnabout could well happen earlier. It is also likely that changes in the composition of municipal wastes, e.g., recovery of plastics for recycling, will precede the projected reduction in total volume. Such changes will alter (reduce, if plastic wastes are separated) the energy content of these wastes.

Agricultural Residues

There is growing evidence that the availability of crop residues has been greatly exaggerated in many studies. Much of this material is consumed locally as part of the production/extraction scheme. If it is diverted, then other energy sources will be required; that is, much of the residue material produced in agricultural operations goes to reduce the external energy demands of these operations.

The projections in Table IV.8 envision an increased availability of agricultural residues until the end of this century, at which time the decreasing supply of chemical fertilizers may be expected to lead to increased local recycling of such materials. These projections are considerably less reliable than those for municipal wastes because:

(1) current estimates on the gross availability of agricultural residues vary widely,
(2) real availability (gross availability less local consumption) has not been accurately determined, and
(3) both real availability and energy costs for collection will be markedly affected by any substantial trend toward low-energy agriculture.

Terrestrial Energy Farms

The technologies required for energy (fuel) farming already exist in agriculture, forestry, and in the biomass conversion techniques currently being practiced (direct combustion, anaerobic digestion). The projections in Table IV.8 (and the published information from which they are drawn) are all based on the use of conventional plant types. The primary energy sources are carbohydrate, especially cellulose. An obvious requirement is the development of plant strains especially suited to this objective. Biomass containing higher fractions of lignin-like components, oils, and hydrocarbons would offer better fuel values than wood and other cellulosic materials.

Several optimistic estimates of the potential for energy farming have been made over the last three years, but these estimates have also been sharply criticized. (See, for example, Caleb [1976].) Three major points are raised against energy farming.

(1) Technical/Economic
 (a) adequate land areas are not available
 (b) water supply is inadequate
 (c) phosphorous requirements cannot be met
(2) Ecological
 (a) soil quality is endangered by large-area, single-crop agriculture
 (b) climatic changes could develop
 (c) energy crops might be vulnerable to crop pests, drought, etc.
(3) Competition with Food Production

Many of the energy farming concepts date from several years ago when crop surpluses were common in the U.S. Now that food has become a major export, and there are prospects of worldwide food shortages, these concepts may not be valid.

Marine Energy Farms

Marine energy farming has been proposed as an alternative to the terrestrial farm. This approach avoids the limitations on water and, possibly, mineral nutrients which have been cited against land farming.

On the other hand, the schemes proposed are largely hypothetical and the technologies unproven. If marine energy farms are to be significant, huge ocean areas will be required. (The projections in Table IV.8 are based on the establishment of 100,000 acres of "farm" by 1985 with a 10 percent annual increase thereafter.) Since few hard data are available, the projections for marine farming are the least reliable of all.

<u>Land/Sea Areas Required</u>. From the data now available, the following area requirements for terrestrial and marine energy farming may be estimated.

Land farm--wood	-- 28 acres per billion Btu thermal energy
Land farm--hydrocarbon (rubber)	-- 27 acres per billion Btu thermal energy
Marine farm	-- 9 acres per billion Btu thermal energy

Costs

The cost for energy recovery from municipal wastes is dependent upon the nature of the refuse feed and related value recoveries (metal, glass, etc.). We assume that there is no charge for the delivery of the municipal wastes to the plant, and that the credit for taking the waste is applied to the cost of disposing of the slag and ash from the plant. The cost of the energy from municipal waste is then a charge on capital and the operating cost of the plant. Based on a plant that processes 275 tons per day of refuse, we estimate the cost of the thermal energy produced to be about $2.10 per million Btu. The major uncertainty in this cost estimate is the credit given for taking the municipal waste and for the recyclable materials that are recovered.

The cost estimates for energy from agricultural wastes are complicated by uncertainties in the cost of waste collection and transportation, the wide variety in composition of the wastes, and uncertainties about the value of the sludge from anaerobic digestion. Our estimate of $2 per million Btu does not include the cost of transporting the fuel to point of use.

A variety of cost estimates for energy farming have been published. They vary from about $0.50 to $2 per million Btu of "raw biomass" energy. If this energy is to be converted to either liquid or gaseous fuels, the conversion cost must be included. Such estimates vary widely, depending on assumed technologies, plant capacities, and conversion efficiencies, but the range encountered runs from about $0.70 to $3.50 per million Btu. This means that, at the high end, a gaseous fuel (methane) produced by bioconversion of solar energy could cost up to $5.50 per million Btu.

SOLAR THERMAL CONVERSION

Solar thermal conversion can potentially serve either for central station generation of electricity, or for distributed generation at the sites of the demands to be served. We distinguish between these two in our estimates of energy contributions and costs. The latter we shall refer to as solar total energy systems, since the primary advantage of distributed generation is that it allows more easily for the reject heat from the thermal cycle to be used for other purposes. There may be some cases of distributed solar thermal generation for local demands without use of the reject heat, but applications will probably be dominated by total energy systems because of their relative economic advantages.

The major effort in government and Electric Power Research Institute (EPRI) programs on central station generation is the development of the central receiver concept. This concept is well suited to units of about 100 MWe rating. It is still possible that alternatives to the central receiver concept (systems with other geometries of mirrors or heat collection) will be found superior for central station generation, but most recent design studies have found in favor of the central receiver concept. With support adequate to pursue only one concept to the pilot plant stage at this time, both ERDA and EPRI have chosen the central receiver (though with different conversion cycles). The preliminary characteristics of the ERDA 10-MWe pilot plant to be built in Barstow, California, are shown in Table IV.9. This concept is probably limited to the southwestern region of the U.S., at least for economic conditions anticipated for the period of time covered by this study. It is planned that the government's central receiver plants will include high-temperature heat storage sufficient for them to act as intermediate load plants. The EPRI design uses distillate fuel as a backup. This would provide for the major part of the daily peak load of utilities in the Southwest. Thus the most probable scenario for deployment of central station solar thermal conversion is a large number of 100-MWe units, of the central receiver design, distributed throughout the southwestern U.S., and providing capacity for the local utilities to match much of their daily intermediate load. As the solar thermal fraction of the generating capacity of these utilities increases beyond the intermediate load requirements, it will become necessary to provide for increased thermal energy storage at the solar thermal plants, or to provide for electrical energy storage somewhere on the net, so that the base load can be partially served.

It is more difficult to predict the system configurations and the applications that will be typical of solar total energy systems. The economics and the institutional aspects of these systems are more complex than for systems that provide only electricity to the utility grid. The type of conversion technology to be used is also uncertain: photovoltaic conversion may be used for the electrical generation cycle as an alternative to boilers and turbines.

Solar total energy systems are generally presumed to be small but could approach the size of present-day central station plants if both the electrical and thermal energy from the plant could be distributed to consumers. It is generally conceded that smaller units will sacrifice something in electrical conversion efficiency and in economies of scale. These sacrifices are at least partially offset by the reduced distances for transporting both electrical and thermal energy. Storage requirements may be more serious than for central station solar power plants, since the rates for backup power can be expected to be high if the average demand decreases sharply without a corresponding decline in peak consumption.

It should also be noted that electrical utilities have historically taken a dim view of total energy systems in this country, although it is a major form of energy distribution in Europe. This may present a major institutional barrier for solar total energy systems.

Our estimates for solar thermal conversion call for improvements in present technology. The most serious problem concerns energy storage in both the solar thermal and solar total energy systems. This problem can be deferred in part if we are willing to depend upon the backup

TABLE IV.9 Preliminary Characteristics of ERDA 10-MWe Solar Central Receiver Power Plant Designs

Characteristic	Honeywell	Martin-Marietta	McDonnell-Douglas
Annual Energy (MW-hr)	4.3×10^4	3.4×10^4	3.6×10^4
Collector Subsystem			
Heliostat Construction*	glass mirror, low-profile steel frame	glass mirror, steel frame	glass mirror, steel frame
Number of Heliostats	2320	1718	2350
Reflective Surface Per Heliostat	40 m^2	37.2 m^2	30.8 m^2
Total Area Reflective Surface	92,800 m^2	63,866 m^2	72,380 m^2
Field Size	308 m radius	565 m × 565 m	527 m × 527 m
Receiver Subsystem			
Receiver Type	vertical cavity	horizontal cavity	external absorber
Tower Height	146 m	137 m	101.4 m
Receiver Working Fluid	water/steam	water/steam	water/steam
Storage Subsystem			
Storage Mechanism	latent heat	sensible heat	sensible heat
Storage Media	salts	HITEC/hydrocarbon heat transfer fluid	rocks/hydrocarbon heat transfer fluid
Electrical Generation Subsystem			
Turbine Rating	15 MW	12.5 MW	15 MW
Turbine Fluid	steam	steam	steam

*A heliostat is a concentrating mirror for central receiver design.

capability of the electric utilities, and assume a willingness either to accept widespread pumped hydro plants or to gamble on finding other storage solutions before the problem becomes serious. High-temperature storage systems are being developed as part of the DOE central receiver pilot plant. Initial experimental results are encouraging.

Solar collectors are the major cost items in both systems. The critical number is the cost per square foot of the concentrating mirror (called a heliostat in the case of the central receiver design). Little effort has yet been expended on this problem and little confidence can be placed in any prediction of future costs. Recently, contracts have been let for concentrating mirrors at a price of \$14 per square foot. Design studies indicate that costs of about \$8 per square foot can be achieved in mass production of heliostats.

The acreage for a given solar thermal conversion plant is critically dependent on location, temperature regime, assumed efficiency of collection and conversion, and time of year. A rule of thumb in the southwestern U.S. is that a square mile will supply about 100 MWe, assuming a 40 percent load factor. However, this slightly understates the required area for a given capacity since the availability of conventional plants is about 70 percent. In solar thermal total energy systems, the possibility exists for using roof areas so no need may exist for the land areas defined above.

The total land area required to meet the U.S. intermediate electrical load demand at the moment (installed 120 GW) is thus found to be about 1,200 square miles, about 1/30 of 1 percent of the U.S. land area.

There have been a number of cost estimates made for the central receiver design. The most important parameter in these estimates is the cost of heliostats in mass production. The capital cost of the plant per rated kW is about \$700 plus \$40 times the cost of the heliostats per square foot. There is a considerable range of cost estimates in the design studies performed thus far, but this relationship is at least approximately correct. It is also estimated that heliostats will cost \$6 to \$8 per square foot in mass production. It has been suggested that advanced designs may reduce this cost to about \$4 per square foot. We use \$8 per square foot in our estimates. Another important parameter that must be specified is the cost of high-temperature energy storage. There is confidence that the cost of sufficient storage for a solar thermal plant to provide electricity at a load factor of 0.5 will be \$200 per rated kW or less. Thus we assume that by the year 2000 the capital cost of a solar thermal plant operating at a 0.5 load factor will be \$1,200 per rated kW. This value holds only for the southwestern U.S.; costs will be higher in other regions.

The costs for solar total energy systems are much harder to estimate because of major uncertainties. First, the type of hardware to be used is not yet known. A variety of design alternatives are now being investigated and it is possible that a number of quite different designs will eventually share the market, each being best for a particular situation. Second, the type of load to be met cannot be well defined. Various applications will require different mixes of electrical and heat energy, and will require heat at different temperatures, or at different times of the day or year. Some of the loads that might be met include space heat, hot water, and electricity for a housing development; air conditioning and electricity for an office building; or process heat and electricity for a factory. Finally, and probably most important as a source of uncertainty, the institutional and financial arrangements that will be used are unknown. We have little institutional experience with total energy systems, and it is unclear who will finance and own them, how charges will be levied for delivered heat or electricity, and what the role will be of utilities that might have to provide backup power.

A detailed estimate of the costs and energy contributions of solar total energy systems would disaggregate the potential market into the various types of loads, estimate capital costs for each type of system, and estimate the financial arrangements that will be made to deduce the cost of delivered energy in each case. However, such complexity is not appropriate to the goal of this report--providing input to national economic models and energy scenarios. Thus we treat this complex market as a single unit and provide a single cost estimate for each year intended to simulate the average of many types of systems being installed at that time. This estimate is very uncertain. Furthermore, to this cost of a low-reliability source of electricity must be added a cost for backup power, the details of which will depend upon the particular application. Studies now under way on the potential applications of total energy systems to residential, commercial, and industrial loads should provide the basis for better estimates in the future.

Similar uncertainties apply to the estimation of the potential energy contributions of solar total energy systems. Here we have used the assumption that the market will be driven by the sales of nonresidential solar air conditioning and solar process heat. This neglects the possibilities for using solar total energy systems to meet residential heat requirements. We would include such contributions if we could see a way to make estimates of how sales will occur, but the institutional complexities (especially those surrounding land-use planning and zoning) make this very difficult. It is our judgment that the values derived from the markets for solar process heat and nonresidential solar air conditioning provide a sufficient surrogate (for the purpose of this report) for a complete specification of the many ways that solar total energy systems might be employed. See Table IV.10.

Solar thermal conversion is often cited as a backup technology that might be rapidly deployed on a crash basis if other energy options are closed (as by a nuclear moratorium and a coal strip-mining moratorium). Thus it is interesting to examine the feasibility of a drastically accelerated deployment schedule for solar thermal conversion. For example, such a schedule might call for converting, in the 10 years between 1985 and 1995, to solar thermal for nearly all new generating capacity. If R&D is pursued aggressively between now and 1985, then an adequate technology for intermediate loads will probably have been demonstrated by that time. (There remains a technical problem of meeting base load demands.) We can then ask if the materials and manpower would be available for the construction of solar thermal plants at such a rate. Recent design studies indicate that there are no crucial requirements for special materials, or for special skilled manpower, that would prevent us from installing solar thermal capacity at the rate we are now installing

TABLE IV.10 Potential Energy Contributions and Costs, Solar Thermal

Solar Thermal Type	1975	1985	1990	2000	2010
Central Station[a]					
Energy Contribution, Quads[b]	--	0.001	0.006	1.0	2.5
Rated Capacity, MW	--	50	1,000	15,000	40,000
Load Factor[c]	0.2	0.2	0.33	0.45	0.45
System Cost, $/Rated kW[d]	7,500	2,500	1,400	1,200	1,200
Cost of Electricity, Mills/kWh[e]	630	210	75	50	50
Solar Total Energy Systems[f]					
Energy Contribution, Quads[b,g]	--	0.001	0.05	0.35	1.6
Heat Produced, Quads[h]	--	(0.002)	(0.09)	(0.60)	(2.7)
Rates Capacity, MW[i]	--	67	2,420	13,400	55,000
Load Factor[j]	0.2	0.2	0.25	0.30	0.33
System Cost, $/Rated kW[k]	5,000	1,500	700	500	500
Cost of Electricity, Mills/kWh[e]	423	127	47	28	26

[a]As distinguished from generation of electricity in solar total energy systems.

[b]Converted from the kWh of electricity produced at 10,000 Btu per kWh.

[c]Load factor is assumed to increase in later years as improved energy storage becomes available.

[d]Cost in 1990 corresponds to estimates made for ERDA in 1975. More recent estimates may be higher. After 1990 our estimate includes additional cost of increased energy storage.

[e]A 15 percent charge on capital is assumed to cover depreciation, cost of money, maintenance, and taxes. Since this cost is for electricity generated onsite at a low load factor, this cost cannot be directly compared to utility bus-bar costs. Rather this cost should be compared to the cost of fuel saved.

[f]This includes the generation of electricity as an ancillary product in the provision of nonresidential air conditioning or process heat.

[g]Includes only the energy delivered as electricity, converted at 10,000 Btu per kWh. Small units will require about 17,000 Btu to generate 1 kWh, but credit can be taken only for displacing 10,000 Btu of primary fuel.

[h]This is the portion of the heat collection for nonresidential air conditioning (70 percent) and the portion of the industrial process heat (50 percent of capacity added after 1985) that is considered appropriate for ancillary production of electricity. This heat should *not* be counted as an energy contribution in addition to the entries in the tables for nonresidential air conditioning and industrial process heat.

[i]Includes only the electrical generating capacity, not the rate of heat energy delivery.

TABLE IV.10 Continued

---

[j]Load factor is assumed to improve as better storage technology becomes available, but to remain lower than central station plants.

[k]Includes only the extra cost above that of the system that collects and distributes the required heat for air conditioning or industrial processes. This cost corresponds to small units (about 10 MWe).

coal and nuclear capacity. There will be larger requirements, per rated kW, for concrete and steel relative to nonsolar plants, but these seem to be within reasonable levels. A much more serious factor, and probably the one which will limit the rate of introduction of solar thermal capacity in such a scenario, is the immense capital requirement that will be required if such a program is attempted before experience with pilot and demonstration plants bring us to a low-cost solar thermal technology. In such a situation, drastic measures for energy conservation, rather than accelerated construction of solar thermal plants, would probably be the more feasible and economical solution.

PHOTOVOLTAIC CONVERSION

The technical feasibility of terrestrial applications for photovoltaic solar energy conversion has been proven. The major remaining obstacle to immediate large-scale use of this approach is the capital cost of the systems, and this focuses current R&D on the problem of cost reduction. Recent efforts at improving the technology have concentrated on the photovoltaic conversion device--the solar cell--since it is the principal cost contributor to the system.

It is important to think of photovoltaic solar energy converters not as "solar cells" but rather as systems which deliver electric power to satisfy the demands of given loads. These loads may range from small single-purpose devices, such as navigation lights, to the peak power demand of a community or an entire utility network.

Photovoltaic systems intended to supply large amounts of electrical power may be categorized as local or central station systems:

(1) Local systems, with the converters attached to or incorporated into structures, and supplying energy to activities connected with structures. Array size, 50-10,000 square meters; average daily output, 25-5,000 kWh.

(2) Central station systems serving a distribution network or a single large consumer. Array size, 1,000 to 50 million square meters; average daily output, 0.5 to 25,000 MWh.

The onsite systems of type (1) or, for a single larger user, of type (2), reduce the need for energy transmission and distribution with the associated losses and costs, but may also involve increases in costs for market distribution and operation and maintenance.

Designing the photovoltaic arrays directly into the building structures may permit some savings in supports and foundations, and offers aesthetic advantages. The basic installation resembles that of a solar thermal collector. One approach is the photovoltaic total energy system which combines the photovoltaic array with a thermal collector. This serves the building's requirements both for thermal energy (for hot water, space heating and cooling, and low-temperature process heat) and for electrical energy (for lighting, motive power, electronics, and high-temperature process heat).

As in most other solar energy systems, a means of energy storage is needed, in this case storage of electrical energy through batteries, electrolysis-cell/fuel-cell systems, flywheels, pressurized gases or pumped hydro storage systems. Also, since it is probably uneconomical to install local storage capacity for more than an average day's energy demand, a source of auxiliary energy will be needed. This source can be based on bulk-fuel delivery with local storage and conversion, or on electric power delivery through the distribution system of an electric utility. The latter would propose new peaking problems, lower the off-peak utilization of generation and distribution capacity, and consequently raise rates unless local on-net storage capacity were available and integrated with the utility so as to effect load leveling.

Present Status

Photovoltaic solar energy conversion devices have been fabricated in pilot-line quantities at various times from silicon (Si), selenium (Se), copper sulfide/cadmium sulfide (Cu_2S/CdS), and gallium arsenide (GaAs). Selenium cells were produced in commercial quantities from the 1920s to the 1960s, while the silicon solar cells have been available from commercial production lines in the U.S. since 1956.

Silicon solar cells are currently commercially available with an average conversion efficiency in space sunlight (airmass zero) of 11-11.5 percent, yielding approximately 150 W per square meter. Their conversion efficiency in terrestrial sunlight (nominally airmass one) is 13-14 percent, yielding 5 to 40 W per square meter on a 24-hour average, depending on location and season. Cells with approximately 16 percent terrestrial conversion efficiency are available in pilot-line quantities ("violet" cells), and developmental cells with about 18 percent nominal terrestrial efficiency can be obtained in laboratory quantities.

Copper sulfide/cadmium sulfide solar cells are made in laboratory and pilot-line quantities in both the U.S. and France. Average efficiencies of 3.5 to 5 percent have been obtained in the U.S. from flexible thin-film devices of 7.6 cm by 7.6 cm (3 in. by 3 in.), encapsulated with plastic.

Developmental gallium arsenide cells have been prepared with efficiencies comparable to those of the recent laboratory silicon cells. Compared to silicon, gallium arsenide allows more favorable high-temperature operation and is, therefore, attractive for use with optical concentration systems.

Research efforts are under way on a wide array of semiconducting materials with potentially favorable properties, including the various binary and tertiary compounds of elements from either the III^d and V^{th} columns of the periodic table, including gallium arsenide (GaAs), indium phosphide (InP), and aluminum antimonide (AlSb), or from the IInd and VIth columns of the table, such as cadmium telluride (CdTe), cadmium sulfide (CdS), and cadmium selenide (CdSe).

Reduction of Cost

The current costs constitute the principal barrier to large-scale terrestrial applications of photovoltaic solar systems. The cost reduction needed is approximately two orders of magnitude, and this will have to be achieved without a significant decrease in conversion efficiency.

Cost reduction can be pursued by several routes: use of raw materials of low cost; reduction of the amount of materials used; and reduction of processing costs (labor, capital costs, and energy). Another approach reduces the solar cell area required to supply a given amount of power through the use of optical concentration; this is economically effective up to the point where the price per unit area for the solar cells falls below that of the optical concentrator with tracking system (correcting for losses in diffuse sky radiation and for optical and other losses).

The basic physical properties of some of the materials suitable for photovoltaic solar energy conversion are such that they inherently require a greater thickness (50 μm or more) of the semiconductor layer as well as greater crystal perfection than do the others in order to achieve comparably high performance. Thus, the materials of the first type will have to be cheap and abundant, as is silicon, in order to fulfill the requirements for large-scale applications. The less abundant and higher cost materials will have to be of the second type, such as GaAs or InP, in order to be able to meet economic constraints. While these materials could conceivably be used with a layer thickness of 1 μm or less, thicker layers may be required due to factors such as interactions with the substrate materials which affect both electrical performance and mechanical integrity.

Silicon solar arrays are currently sold for \$13 to \$20 per W of *peak* output, or approximately \$1,300 to \$2,000 per square meter, in quantities of a few hundred square meters. Where Cu_2S/CdS cells have been or are being sold from pilot lines, their prices are close to those of the silicon cells. GaAs cells are demanding significantly higher prices and are considered primarily for use with high-ratio optical concentrators, the economies of which are still quite unclear. Packaged power systems, including storage batteries, deliver about 15-22 W per square meter continuously, depending upon the site. To deliver electrical energy competitively for large-scale application, the price of the arrays will have to fall at least to the \$10-\$20 per square meter range for high efficiency (>10 percent) arrays.

At that level, solar photovoltaic power can be competitive in a wide range of applications. Since photovoltaic systems incur no direct fuel costs (except for required backup power) and other operating costs should be small, the capital costs predominate. The energy costs are thus determined primarily by first cost, system life, and interest rate. First cost includes energy storage, and power conditioning, such as regulation, inversion, etc. Potential credits due to integration or siting advantages, such as savings in roofing materials, combination with thermal solar energy utilization systems, and reduction (by local siting) of the usually considerable transmission and distribution costs for electric power, can provide potential economic advantages (if the problem of providing backup power at low cost can be resolved).

Future Developments

In studying the plans that have emerged during the last years regarding government stimulation of the development of low-cost photovoltaic systems for large-scale terrestrial applications, one gains the impression that this development will occur in essentially three phases. In Phase I, until 1985, primary emphasis will be placed on early systems demonstration. During this phase, ERDA expects that solar arrays of 300-MW peak capacity will be installed, including 10 MW to be purchased by ERDA, and 215 MW by other U.S. government agencies, for demonstration projects. Some of these applications at remote sites may be cost effective even at 1975-85 prices. These large government purchases are meant to provide a subsidy to enable industry to establish a considerable production capacity (4 square kilometers per year) and to develop the terrestrial market. Concurrent with obtaining production experience, data will be gained with operation of the demonstration and application systems. Because of the production emphasis in Phase I, an early transition to pilot-line operation and soon thereafter to full-scale manufacturing will be needed. Thus, this phase will be dominated by the application of those processes and technologies which are fairly well understood at the current time. It should be noted that considerable use may be made during Phase I of optical concentration up to moderate concentration ratios (<20). Such use of concentrators may be required to reach the ERDA cost goal (50¢ per W peak) set for the end of Phase I, although the array costs themselves may not be able to reach this desired level. In Phase II, however, optical concentration should start to appear less attractive because of the sought-after reductions in silicon solar array prices.

In Phase I, the reduction of price (from about $20 per W peak now to 50¢ per W peak in 1985) is assumed to come about through production experience and the development of partial automation and bulk processing for array production. This assumption, fundamental to the government program in photovoltaics, is the most drastic example in the solar R&D programs of the anticipation of learning-curve effects. The reduction in cost sought is a factor of 40. Purchases of 3 square kilometers of arrays with a peak output of 300 MW, at a total cost of $500 to $1,500 million, may be necessary to drive the learning curve to this point. If the learning-curve effect is indeed as strong as assumed, then the majority of the purchases, which occur in the later years, will be about 50¢ per W peak. However, if the learning-curve effect has less slope, then the cost in 1985 will be higher, and the total cost of the program of purchases will be very high. In order to induce industry to invest in automated (or at least semiautomated) production facilities, the government must assure the producers of a guaranteed minimum of sales. If the desired cost reductions do not occur, then the government must pay an expensive bill.

The dependence upon the learning curve for gaining a factor of 40 in cost has been criticized by some outside the government who argue that advances in technology will come about at a lower cost through direct funding of advanced R&D. Currently the direct funding for R&D in the photovoltaic program is primarily for support of the improved production technology required for the 1985 goal. This situation does not meet with favor from the critics who view the Phase I effect as an expensive step, unnecessary for the ultimate development of the advanced technologies that all agree will be needed to meet the year 2000 goals. The government views the development of an industrial capacity for the Phase I technology to be an essential step for the later rapid deployment of advanced technology.

Phase II, which will extend to approximately 1995, will be characterized by the transition from processes based essentially on current technology, to the ultimate high-speed and low-cost processes which are to be based on a new technology. The roots of this new technology are starting to appear now. Thus, the beginning of the second phase cannot be pegged to the end of Phase I, since considerable research and development that will form the basis of the developments of Phase II will be occurring concurrently with the first phase. Phase II is the period when new technologies in compound semiconductors and polycrystal devices may acquire a significant share of production.

Finally, Phase III will see a maturing industry with more slowly increasing production rates and a steady refinement of methods and processes.

The learning curve provides a tool for the assessment of the goals of such a program. The availability of 300 MW of solar arrays in 1985 requires somewhat more than an annual doubling of the production rates. The learning curves for silicon solar cells (Figure IV.5) indicate a 1985 price of approximately \$100 per square meter, or 80¢ per W peak, if the perhaps somewhat optimistic 70 percent learning curve is followed to the required cumulative production of 3 square kilometers. Using the 85 percent learning curve as the upper limit for a successful development program, a price of \$600 per square meter would be obtained, corresponding to \$5 per W peak. In the second phase, this price range should be reduced to \$21-100 per square meter, corresponding to 15-70¢ per W peak. The open-ended Phase III would include gradual further cost reductions, perhaps to the level of \$16-50 per square meter or 10-30¢ per W peak.

When the learning curve for a product goes through several orders of magnitude in cost, it is usually through a sequence of technologies, rather than through drastic improvement of a single technology. Thus it is generally assumed that as the improved production based on a present technology reaches its limit of perfectability, in about 1985, a new technology based on advanced techniques or materials will emerge. The process of perfection of this advanced technology during the period 1985 through 1995 will then allow prices to continue to decrease.

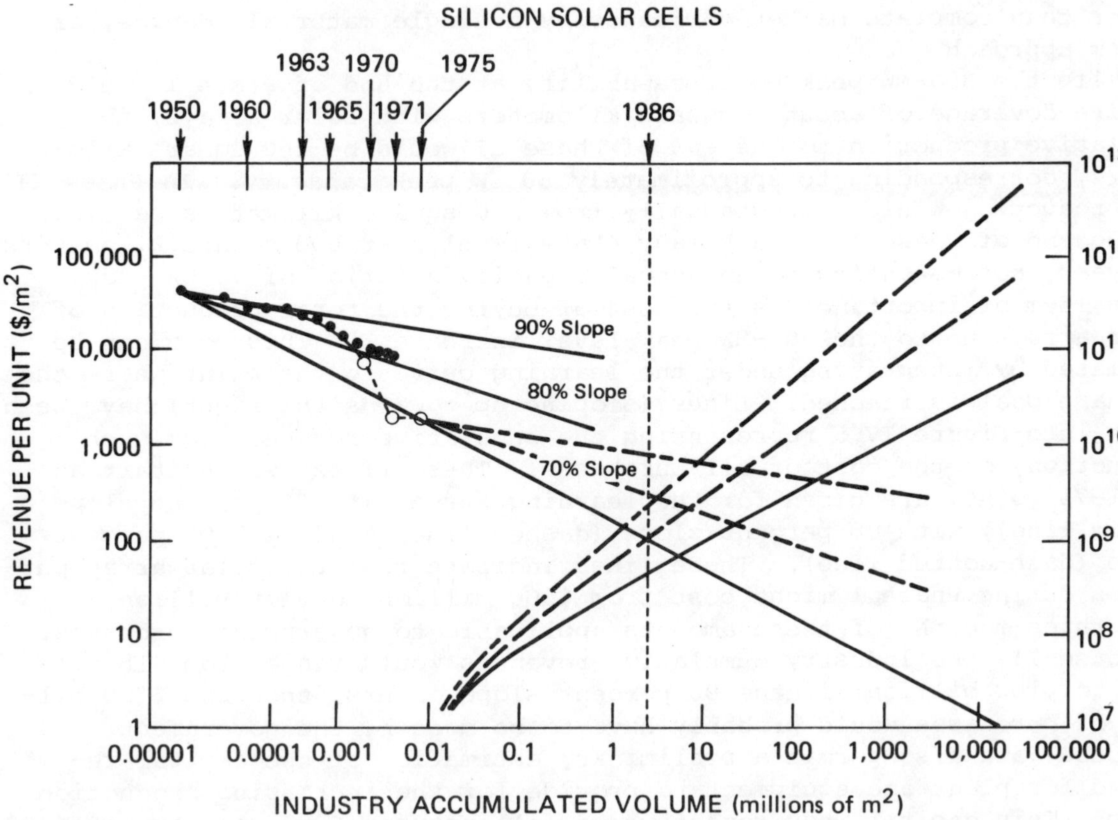

FIGURE IV.5 Learning curves for photovoltaic arrays. NOTE: The scale at the right shows the cumulative revenue of the photovoltaic industry in each of the three scenarios of technology development.

Since the learning curves are gross statistical tools, including such effects as technology changes, they should be equally applicable or inapplicable to photovoltaic systems of any type. One material may completely replace another during the market development represented by the learning curve. Similarly, there may be transitions between nonconcentrating and low-ratio or high-ratio optical concentrator systems. All of these would be considered technology changes taking place on the generic photovoltaic array learning curve. This generic learning curve is assumed to be a composite of "type" learning curves, which describe individual device or array types, the latter including, for instance, optical concentrators. A type learning curve may slow down in its progression or level out completely, or come to an end as another more economical type replaces the first in the market. Throughout this process, the generic learning curve continues its progress.

The discussions related to Figure IV.5 may thus be applicable to the compound semiconductors, while future market sharing between the different devices is still fully unknown. By analogy to the historical development of the energy economy, one should expect a certain sharing of the market rather than complete market dominance by a single material, device, or system approach.

While the 300-MW peak power capability at the end of Phase I would require coverage of about 3 square kilometers with solar arrays, the cumulative production to the end of Phase II would be 460 square kilometers, corresponding to approximately 50 GW peak capacity. In Phase III, the production would grow gradually from 200 square kilometers per year at the end of Phase II to a steady state level near 500 square kilometers per year, corresponding to an annual capacity addition of 75 GW peak.

An item of importance is the cost of buying the total production of solar arrays up to the 300-MW peak level in Phase I. This cost can be estimated by integrating under the learning curve to the point where the purchase goal is reached. Lines (sloping up towards the right) have been drawn into Figure IV.5 representing the cumulative revenue from such production, or the cost of all purchases. These lines, which start at the 1975 point, are given for the learning curve with 70 percent slope (solid line), with 80 percent slope (dashed line), and with 90 percent slope (dash-dotted line). These lines indicate that the solar array purchases during Phase I might cost from \$500 million to \$1.5 billion, with three-fourths of these amounts applicable to government purchases. In Phase II, the industry cumulative revenues would range from \$15 billion to \$100 billion; if the 90 percent slope occurs, then the \$100 billion of purchases would probably have to be made by the government.

These data also permit a preliminary estimation of the capital required for plant and equipment to provide for the increasing production rates. This capital requirement would fall between \$200 million and \$800 million during Phase I. In Phase II, the required capital investment may amount to \$5 billion to \$25 billion. Finally, in Phase III, the new investment requirements for plant and equipment may slowly decrease and be supplied from depreciation funds. The experiences of Phase I, both with the production processes and the installed systems, and the status of the development of new technologies will determine the potential further significant cost reductions. Although certain price ranges may be derived from the formalism of the learning curves, the lack of information about the technical advances required makes a high-confidence estimation of the potential success of the program impossible. The new information to be gained during the next 9 years should alter this situation drastically before Phase II is started. This assumes that R&D of the new technologies will be adequately funded in the period 1975-1985. The fulfillment of this condition, however, is uncertain; in the view of some observers, the present budget plans, which are oriented towards meeting the Phase I goals by 1985, provide for inadequate research to prepare for the needs of Phase II.

The introduction scenario summarized above can be considered a likely one, assuming normal economic conditions, sustained government support, and continued absence of the appearance of technology barriers. In case of an emergency situation, production buildup could be greatly accelerated through a concerted effort with appropriate funding, although at a cost. Conversely, neglect of the field and reduced government support will probably result in withdrawal of industry and a halt in development.

Contributions and Costs

An energy impact schedule is shown in Table IV.11. The "Annual Energy Delivery" is based on the estimated "Installed Capacity" and the available annual electrical energy output from a fixed-orientation, south-tilted unit array area in the average location in the U.S. An arid area location (southwestern U.S.) could yield approximately 50 percent higher energy output, and an array that tracks the sun, up to 40 percent higher energy output. Similarly, mismatch between solar energy availability and demand statistics, and passing a sizable portion of the array output through a storage system of limited efficiency can result in up to 50 percent reduced energy delivery, as can placement of the array in less favorable climates. The given numbers may thus represent average values for the expected annual energy delivery.

The "Array Capital Costs" of Table IV.11 are the total costs of all installed arrays up to the given point in time. They are obtained from the "cumulative revenue" curves derived from the learning curves. There are three sets of costs given in each column, separated by hyphens. The highest costs are taken from the dash-dotted line in Figure IV.5, and are chosen as the highest plausible on the assumption that photovoltaic system introduction will not be pursued if at least this much price reduction cannot be achieved. The middle values are based on a learning curve of initially 75 percent slope (up to 1990), which then tapers off to a 78 percent slope in the year 2000 to 2010 range (dashed curves in Figure IV.5). This curve reaches the $15.00 per square meter ($0.095 per W peak) value in 2010. For 1985, the array cost value has been decreased by assumption of the use of low-ratio optical concentration.

For the lowest values, the government's goal of $0.50 per W peak in 1985 has been used. Although this value lies significantly below the 70 percent slope learning curve which probably represents the lowest plausible longer range expectations, temporary deviations are possible and it is conceivable that a concerted effort could yield such a value, particularly with use of low-ratio optical concentrators. For the longer range (1990), a return to the 70 percent slope learning curve as the most optimistic value has been assumed. Finally, it will be unlikely that any kind of solar array can be fabricated for less than $10 per square meter. This has, therefore, been used as the long-range lower limit. Consequently, use of the "cumulative revenue" curve ends where the learning curve reaches the $10 per square meter value, with the additional capital cost based on the product of quantity and price, rather than on an integral with varying price.

The estimated costs of the storage subsystem, the inversion and regulation subsystem, transportation and commerical distribution, and system installation are given in footnote *b* of Table IV.11. The latter category, transportation, distribution, and installation costs, is difficult to predict at this time, due to lack of precedent. However, the recent experiences with solar heating systems indicate that these costs will likely be larger than previously thought, and that they may well dominate the system costs in the distant future.

The cost of capital, detailed in footnote *a* of Table IV.11, has, for simplicity, been used as a constant annual rate. It will differ for the

TABLE IV.11 Energy Delivery and Energy Cost for Photovoltaic Conversion

Year	Installed Capacity: km^2	Installed Capacity: MW Peak	Annual Energy Delivery (Million MWh)	Capital Costs[a]: Array Only (Billion $)	Capital Costs[a]: Total[b] (Billion $)	Energy Cost (Mills/kWh)
1975	0	0	0	0	0	--
1985	3	300	0.438	0.3-0.56-1.5	0.35-0.7-1.9	80-190-800
1990	37	4,500	6.57	1.9-3.6-13	2.5-5.3-18	39-105-500
2000	2,200	335,000	489	24-78-280	57-180-510	12-57-190
2010	7,200	1,100,000	1,606	74-150-670	180-470-1,300	11-51-150
After 2010	Each New km^2	160	0.223	0.015	0.06	46

[a] Cost of capital annually 25 percent for industry, 18.5 percent for utilities, 10 percent for homeowners. There may be tax incentives or other subsidies to lower cost of capital for solar energy systems. Applied cost of capital should be a mix resulting from systems use by the three categories simultaneously. Thus, low cost of capital is 10 percent, high 18.5 percent. Probable intermediate value used is 12 percent in 1985, when subsidies are likely, increasing to 13 percent in 1990, 15.5 percent in 2000, and 17.5 percent by 2010.

[b] Including storage, inverter/regulator, and other costs.

Storage costs: up to two-thirds of output delivered from storage at 50 percent depth of discharge, with as little as 60 percent storage efficiency at up to $40 per kWh storage cost.

Inverter/regulator cost: $30 per kW.

Combined costs: $15 to $24 per m^2.

Encapsulation/distribution/installation cost: $30/$m^2$ minimum; maximum $120/$m^2$ in 1985, $100/$m^2$ in 1990, $80/$m^2$ in 2000, and $60/$m^2$ in 2010.

Total estimated costs:

Intermediate: $45/$m^2$ (either no storage or minimum encapsulation/installation cost)

Maximum: $144/$m^2$ in 1985; $124/$m^2$ in 1990; $104/$m^2$ in 2000; $84/$m^2$ in 2010.

Minimum: $15/$m^2$ (advanced storage, encapsulation/installation costs borne by thermal part of total energy system; or construction credits from building integral design cancel part of encapsulation/installation cost.)

various user categories due to differences in depreciation times used, tax treatment, profit requirements, etc. Use incentives provided by government will have a strong influence here, be they in form of subsidies or of tax relief. The composite cost of capital, including all user groups, has been gradually increased with time to reflect a gradual removal of incentives as well as an increasing share of new installations by industry and utilities with their higher costs of capital.

Operating costs, being primarily maintenance costs, have been neglected as insignificant in comparison to the capital costs and in view of the general uncertainty of the data. Obtaining such low operating costs will be an important aspect in system design.

It should be noted that the energy costs given in the right-hand column of Table IV.11 are based on the composite of all systems installed at various times up to the cutoff point, which means that they were bought at gradually decreasing prices. This energy cost is, therefore, higher than the cost of energy from the youngest system installed before the cutoff point. The energy cost of the youngest system at the year 2010 is shown in the last line, based on the most probable intermediate data used in the other calculations for that year. The corresponding value for the year 1985, but based on the lowest plausible data for that year, is 53 mills per kWh. See Table IV.12.

Problems Requiring Solution

Technology development will be required to a substantial degree in solar array processing, from the raw material to the finished array, and in energy storage. Expansion on current technology will not be sufficient to reach the required cost reductions.

Photovoltaic solar energy conversion systems require, with few exceptions, storage of electrical energy. This energy is generally converted for storage into another energy form, and then converted back to electrical energy. Examples are the battery which stores chemical energy and pumped hydropower or the flywheel which stores mechanical energy, with conversion by an electrical motor and a generator. To obtain the desired economies for photovoltaic solar energy systems, considerable technology development on storage devices will be needed. The requirements on these devices differ significantly from those currently under development for other applications. High energy density (low mass and volume), and high discharge rates are not required a priori, although the energy density can influence the storage economy. In contrast, high storage efficiency, long chronological life (20 years) and cycle life (7,000 deep-discharge cycles) are most important for achieving the necessary storage economies.

The current government program emphasizes quick demonstration projects (Phase I), including establishment of significant production facilities which could very soon be obsolete because of newly developed technologies. The latter will be needed to achieve cost reduction to the required level. Needed is a significantly larger research program to acquire the new material and process knowledge required for the technology development of Phase II. The government has not established a broadly based research

TABLE IV.12 Potential Energy Contributions and Costs, Photovoltaic

Level of Technology	1975	1985	1990	2000	2010
Improved Technology[a]					
Energy Contribution, Quads[b]	--	0.004	0.07	4.0	4.9
Rated Capacity, MW	--	300	4,500	33,500	41,000
Load Factor	0.17	0.17	0.17	0.17	0.17
Middle Estimate of Cost[c]					
System Cost, $/Rated kW[d]	20,000	2,467	1,178	1,178	1,178
Cost of Electricity, Mills/kWh[e]	1,990	146	117	117	117
Highest Plausible Cost[f]					
System Cost, $/Rated kW[d]	20,000	6,300	4,000	4,000	4,000
Cost of Electricity, Mills/kWh[e]	1,990	627	400	400	400
R&D Goal for Costs[g]					
System Cost, $/Rated kW[d]	20,000	1,167	556	556	556
Cost of Electricity, Mills/kWh[e]	1,990	116	55	55	55
Advanced Technology[h]					
Energy Contribution, Quads[b]	--	--	--	0.9	16.0
Rated Capacity, MW	--	--	--	60,000	1.1×10^6
Load Factor	--	--	--	0.17	0.17
System Cost, $/Rated kW[d]	--	--	--	537	427
Cost of Electricity, Mills/kWh[e]	--	--	--	53	42

[a] Technology improvement resulting from the quick-response effort under the current ERDA 1985 goals. This technology will be based primarily on currently existing knowledge, stimulated by development support and by government purchases. If an advanced technology (footnote *h*) appears in 1985, it will probably drive the improved technology from the market by 1990.

[b] Converted from kWh at 10,000 Btu per kWh.

[c] Corresponding to a slope of the learning curve of approximately 75 percent until 1990 and 78 percent until 2010.

[d] Including cost of storage, inverter, and regulator.

[e] A 15 percent charge on capital is assumed to cover depreciation, financing, maintenance, and taxes.

[f] Corresponding to a slope of the learning curve of approximately 85 percent. This is the highest cost at which a photovoltaic program might continue to be pursued.

[g] Corresponding to a slope of the learning curve of less than 70 percent until 1990 and then 70 percent until $10 per square meter is reached.

[h] Postulated new silicon process technology or new compound semiconductor materials and technology; and possibly application of concentrators. Fine single crystals or very coarse polycrystalline materials may be used. This new technology will be based on new knowledge gained through R&D carried out in the 1975 to 1985 period.

program on photovoltaics. University groups are only minimally engaged in the program, and usually not in a way which would match their capabilities in advanced research.

WIND ENERGY CONVERSION

In estimating the potential energy contributions and possible costs of energy from wind, it is much more important than in other solar technologies to consider the total size and the geographical distribution of the resource. This is because the local wind energy resource varies by factors of 10 or more between geographical locations (even between locations close together), and because the best wind sites are often not close to points of energy demand. Even the total wind resource of the U.S. is not well known, though estimates have been made by various authors. For example, Reed (1974) has estimated the power available inside the contour of 150 W per square meter--at approximately 13 mph average windspeed--on the mid-continent high plains. He found that several times the present U.S. electricity consumption could be extracted from the winds in this region alone. Coty and Dubey (1976) estimated that 1 trillion kWh of electricity could be produced annually from wind generators in high-wind areas located on land not currently in use or reserved for other purposes. This is about half the electric energy consumed annually in the U.S. today. There seems little doubt that the total amount of wind energy over the U.S. land mass is quite large. Even if all the locations with mean windspeed below 12 mph (5.3 meters per second) are excluded, the total energy is still large.

Just as there is no doubt that the wind energy resource exists, there is also no doubt that the technological means already exist to convert it into electricity. Wind turbines are fairly simple mechanical devices, certainly less complex than an automobile. Their future introduction is not likely to be limited by the rate at which they can be manufactured. The problems of connecting wind generators to an electric grid have been successfully resolved in experiences both in Europe and in the U.S. The period of greatest use was in Denmark during World War II, when fuel supplies were severely interrupted. During that period 88 wind-driven generating plants, of various sizes up to 70 kW, were pressed into service. They generated 18 million kWh for local networks in 7-1/2 years. In terms of present U.S. requirements, this is not large, but it does indicate that networks can use the electricity from wind-driven generators for at least some fraction of their generation.

Given that the resource is large, and that an adequate conversion technology has been demonstrated, the size of the energy contributions that we will obtain from wind energy depends upon the cost of the energy supplied from wind for its various possible applications. The future cost of electricity (or some other form of energy) extracted from the wind and delivered to users is uncertain, and varies over a wide range (perhaps a factor of 5 or 10) depending upon the following factors:
(1) Mean windspeed. The energy in the wind at any instant is proportional to the cube of the windspeed. It costs little more to erect a wind

turbine at a point with 20 mph average wind than at one with 10 mph average; the cost of the power, however, may differ by a factor of 8 in the two cases. For this reason it makes sense to exclude locations and regions with mean windspeed below 12 mph, even though there is no technological barrier to recovering the energy at these locations; (2) diurnal and annual wind variation; (3) accessibility of the site to existing roads and transmission lines; (4) existing generation capacity in the utility system to which the wind generators will be connected; (5) existence of local interruptible loads to which the wind energy may be applied; (6) daily and annual load schedule in the utility or region; (7) the scale at which wind generator units are produced (individual units are anticipated to be about 1 MW rating); and, finally, (8) the development of wind-generator technology.

The cost per unit of energy depends on all the variables noted above. Many estimates of this cost have been made. For example, Hewson estimates that 4,000 units of 1 MW each could be built and installed in 1980 for \$600-\$700 per installed kilowatt (Hewson, 1975). A 1-MW unit, properly matched to its site, will deliver about 400 × (8,700) = 3.5 million kWh per year (40 percent plant capacity factor). With a 15 percent charge on capital to cover costs, including depreciation, cost of money, maintenance, and taxes, this gives 26 to 30 mills per kWh, assuming that all the energy available from the wind machine can be used. (This last point is very critical. See discussion below on the value of energy from an intermittent source.) Wind-generator lifetime of 30 to 40 years is assumed; there is reason to believe this might be realistic. Coty and Dubey give values from 21 to 25 mills per kWh for the fuel cost at which wind generators "break even" when used on a substantial scale. Their estimate drops to 12 mills if hydroelectric peaking energy is available in the system.

A commercial wind-generator manufacturer recently stated that his company would be willing to sell for \$175,000 copies of the 0.2-MW machine they are presently installing for an electric utility in Canada " . . . and the price would be less if you bought more than one."*
This works out to 37 mills or less per kWh on the basis of a 15 percent annual charge on capital and a favorable wind site. Moreover it represents a real-time price, not a projection.

Land requirements will not be excessive. The Lockheed group estimates that 1 trillion kWh per year could be produced from 20,000 square miles in areas not used for farms, parks, highways, or national defense reserves. Only about 6 percent of this, or 1,200 square miles of very low value land, would actually be required for the installations; the remainder is only to insure sufficient spacing for the wind turbines, in order that they not interfere with each other (Coty and Dubey, 1976).

*Lynn Maile, Dominion Aluminum Fabricating, Ltd., Toronto. Mr. Maile made the remark in response to a question from the floor at the Vertical Axis Wind Turbine Technology Workshop, Albuquerque, May 18-20, 1976. (The question was asked by L. Divone, Chief of the ERDA Wind Energy Program.)

The prospects of capacity displacement by wind-generation facilities are not very good. Wind generators are primarily energy sources, not power stations. However, the importance of fuel saving will increase, and that of capacity displacement will diminish, if fuel prices increase relative to the cost of other goods (especially labor).

The Value of Energy from an Intermittent Source

To recover full value from an investment in wind turbines it is necessary to make some accommodation to the intermittent nature of the source. One approach is to couple the wind generators to interruptible loads, such as water pumping, sewage aeration, cooling or heating of large thermal masses, or production of chemicals (including hydrogen by electrolysis). If a connection to the utility grid is used as a backup, the wind-generated electricity can then be valued only at the average fuel cost of the system at the time that cost is lowest (usually in the middle of the night, when only the most efficient plants are operating). That cost today is certainly less than 10 mills per kWh; it could be less than 5 mills. This argument assumes that the wind generators and their loads are a small fraction of the capacity and load of the electricity utility, and that the utility is in fact connected as a backup. There are several approaches to the direct use of wind generators in electric utility networks:

(1) Storage of some sort may be provided for each wind generator, or each cluster of generators, or for the entire wind component of the system. In this case all the wind-generated energy is stored until needed. If delivered at a time of peak electricity demand the wind-generated electricity can be valued at the fuel cost of the peaking generators--about 30 mills per kWh. Storage large enough to absorb the output of the wind generators for 6 to 12 hours is required. At present the usual and established way to store energy on this scale is with a hydroelectric installation. In a grid with existing hydro capacity the storage can be accomplished by displacement; that is, simply by not using as much water to generate electricity at times when the wind generators are delivering power. The water not used is held behind the dam for future use. The large Smith-Putnam wind turbine, connected to a system with abundant hydropower, was operated in this manner. However, displacement storage cannot be used unless the network already gets a large portion of its energy from hydro, since otherwise the hydro capacity will likely already be scheduled as peaking power. In this case, when the wind generators deliver energy in the middle of the night there is nothing to displace. Systems with installed pumped storage, rather than ordinary hydro, may be able to make good use of wind energy if sufficient pump capacity is available, since wind energy can then be converted into lifted water without disturbing the remaining part of the network. Systems with ordinary hydro could handle wind energy in the same way if separate pumping loops (to lift water from the lower pool to the upper reservoir when wind energy was available) were installed. Of course, this entails a cost, while displacement storage costs nothing.

Another type of large-scale storage suited to use with wind generators is pneumatic--compressed air in underground caverns. This is not yet a well-established and standard technology, but at least one large storage facility has been constructed in Europe. Other possibilities, such as flywheel storage at the substation level, are still in the future.

(2) Without storage, it may be argued that a large number of wind generators, goegraphically dispersed, will still displace a certain amount of system capacity because the wind is not likely to disappear in all spots simultaneously. Whether this is sufficiently certain to satisfy utility reliability requirements is not an easy question to answer. If a utility has at least a small amount of easily available, low-cost peaking capacity, such as hydro, the prospects for using wind without storage are greatly improved.

(3) If the wind-generated electricity in the grid cannot be regarded as reliable then it may not be usable at all during off-peak periods. This is because base load power plants, producing electricity at low fuel costs, are typically huge, and cannot be turned on and off rapidly. Response times vary; perhaps half an hour is typical. Peaking plants have response times of only a minute or two, so they can be used with wind generators rather easily, but in the middle of the night the peaking plants are not usually operating. In order to be able to make use of wind-generated electricity to save fuel in the base load plants, Coty and Dubey have proposed "grid stability storage." (They propose electrochemical storage batteries, but other forms are possible.) The idea here is to provide a small amount of storage--enough to absorb wind generation for perhaps a half hour. If the wind generators are carrying part of the system base load and the wind dies, the grid stability storage provides the needed energy while the conventional (coal or nuclear) capacity gets operating. If the wind returns, the storage is recharged while conventional capacity is being shut down. Coty and Dubey showed that the amount of wind energy which could be employed in a typical grid was seven times greater with grid stability storage than without it. Considering the low fuel cost of base load power, however, and the rather considerable cost of storage, it remains to be established that this idea is worthwhile.

Summing up, it is unfortunately true that fuel saving is much less valuable than capacity displacement. If it could be arranged that the wind would be turned on only at times of peak demand, then wind-generated electricity could fairly be said to be worth about 30 mills per kWh, since this is now the fuel cost of electricity produced by the gas turbines and diesel engines used for peaking. The fuel cost of electricity produced by large coal and nuclear base load plants is, however, much less. Wind-generated electricity produced in the middle of the night has very little value, unless it can in some way be stored for use in times of high demand the next day (thus realizing its full fuel-saving potential) or, alternatively, unless the wind-generated part of the total installed capacity of the grid is sufficiently reliable that it can carry some of the base load. In this case, some of the coal or nuclear capacity need not be built and the wind-generated energy has a value comparable to the total cost of electricity from the coal or nuclear plants, instead of just the fuel cost.

Considering the dominant role of unknown and unpredictable factors, nearly all of them nontechnical, the schedule in Table IV.13 amounts to little more than a guess. The maximum possible rate at which wind-electric

TABLE IV.13 Potential Energy Contributions and Costs, Wind Energy

Load Type	1975	1985	1990	2000	2010
Interruptible Loads[a]					
Energy Contribution, Quads[b]	--	0.1	0.5	1.4	1.8
Rated Capacity, MW[c]	--	2,850	14,250	40,000	50,000
Load Factor[d]	0.4	0.4	0.4	0.4	0.4
System Cost, $/Rated kW[e]	875	700	537	513	490
Cost of Electricity, Mills/kWh[f]	37	30	23	22	21
Contribution on Other Loads[g]					
Energy Contribution, Quads[b]	--	--	--	3.6	19
Rated Capacity, GW[c]	--	--	--	100	520
Load Factor[h]	0.2	0.2	0.2	0.2	0.15
System Cost, $/Rated kW[i]	875	700	537	800	1,200
Cost of Electricity, Mills/kWh[f]	74	60	46	68	137

[a] Water pumping, production of chemicals, various other onsite applications, and generation of electricity on utility grids that have existing hydro capacity that can be used for peaking. It is assumed that such applications are limited to 5 percent of the nation's total generating capacity.

[b] Converted from kWh at a rate of 10,000 Btu per kWh.

[c] Electrical output at rated wind speed of each system.

[d] A high load factor is assumed corresponding to use of the best wind sites and applications that can always use wind-generated electricity.

[e] The 1975 cost corresponds to an estimate for production of 0.2 MW units of the type now being installed in Canada. Decreasing costs with time are assumed to be due to learning curve.

[f] A 15 percent charge on capital is assumed to cover costs, including depreciation, financing, maintenance, and taxes.

[g] All applications that cannot be run in a wind-following mode.

[h] The load factor is assumed until 2000 to be half that of systems serving interruptible loads because of lack of match to time of peak demand. Decreased load factor in 2010 is assumed to be due to use of poorer wind sites and poor matching to loads.

[i] Increased cost per rated kW in 2000-2010 is assumed to be due to the use of sites with lower average wind speeds because best sites have already been used.

generators could be installed is considerably higher. For example, Coty and Dubey project 1.2 trillion kWh per year by 1995, an annual displacement of 12 quads of primary energy.

The Table IV.13 schedule requires only improvements of present technology. Or, perhaps one should say, present technology applied to build new and improved machines. In particular, the development of the Darrius rotor and other innovative technologies should be pursued, but the scenario described does not depend on these new types of devices. Use of wind energy is not a technology-limited problem. New storage technologies would likely have a greater impact than new wind turbine technologies.

Without doubt, progress in the government wind program would benefit from an increase in funding. Funding to date has been appropriate to an R&D effort, but now that the program is ready to move into larger scale demonstration and development efforts the cost of the program will increase. With more funding the government could do the following: (1) Against broadly drawn specifications, let contracts for production of a dozen or so 1 MW units to each of five or so manufacturers. This would get the actual design effort to industries accustomed to cost-effective design and production. It would also get more units up faster, so that problems would become apparent sooner. (2) Fund a large number (perhaps 100) of mock cluster installations of 20 or so wind generators each. The wind generators would be represented by anemometers, and the computed energy output would be matched against the real-time load on the utility system to which generators would (if built in future) be tied. (3) Subsidize production costs to let the market develop to the point where scale production economies develop.

The present government effort will result in two 0.2-MW units on utility grids late in 1977 and two 1-MW units on utility grids about two years later. This is a rational program and a good use of the available funding, but it means that by 1980 it will still not be possible to move forward on a commercial basis. Interfacing problems will still be appearing and no substantial production capacity will exist.

OCEAN THERMAL ENERGY

Ocean thermal energy conversion (OTEC) is a solar energy scheme to produce large quantities of power from the thermal energy stored in the tropical oceans. It is unusual among solar energy technologies in that it requires no energy storage and can operate day and night throughout the year. Power must be used at the site or transmitted via underwater cable from the plant site to the shore with further transmission inland through conventional power lines. The sites available to the United States are: the Gulf of Mexico, the Gulf Stream, the Hawaiian Islands, the Virgin Islands, Puerto Rico, and the open seas. Energy-intensive manufacturing plants might be located at the OTEC sites to avoid transmission problems and cost. In our estimates of energy contributions and costs we treat such uses as being displacement of other on-grid generation of electricity.

Early OTEC plants are envisioned near the various islands mentioned to feed energy directly into existing power grids. Subsequent development calls for numerous large (400 MW) OTEC plants located in the Gulf of Mexico, off the Florida, Alabama, and Louisiana coasts, and in the Atlantic near Florida, Georgia, and the Carolinas.

Cost estimates for ocean thermal plants have been the subject of sharp disagreement. Early estimates showed quite low construction costs, but these studies have been identified with the views of active proponents of OTEC. There is unfortunately little direct experience on which to base good estimates of the cost of constructing such facilities and keeping them operating in an ocean environment. However, there is some indirect experience based on other ocean technologies. In 1975, TRW and Lockheed completed two one-year studies intended to evaluate the claims of earlier OTEC proponents, to incorporate the knowledge of companies experienced in ocean technology, and to evolve baseline system designs using state-of-the-practice components (Lockheed Missile, 1975). The two teams independently concluded:

(1) OTEC is technically feasible.

(2) The cost of the heat exchangers is 50 to 60 percent of the cost of an OTEC plant based on present technology.

(3) The plant capacity factor is in the 80-90 percent range.

(4) The plant capital cost with 1975 technology and dollars is $2,100-$2,600 per kW net output.

(5) Aluminum holds the greatest prospect for future development of an advanced technology to reduce cost of the OTEC heat exchangers. (Titanium was used in the baseline design of the heat exchangers.)

(6) Improvement in heat transfer technology has a great potential payoff.

(7) The cost of bus-bar electricity ranges from 30 to 72 mills per kWh, depending on the various assumptions concerning the site, design, and method of financing. This does not include cost of transmission to shore.

The cost of the heat exchangers deserves special discussion. Because of the small range of working temperatures, the thermal efficiency of an OTEC plant is necessarily low and, therefore, the amount of heat to be exchanged necessarily large and the heat exchangers themselves large. At 2 percent efficiency, approximately 340,000 Btu per hour need to be exchanged to generate 1 kWh (170,000 Btu into the evaporator, 170,000 Btu back out in the condenser). Typical heat exchangers in marine service, when well maintained, will transfer between 300 to 450 Btu per hour per square foot per degree Fahrenheit. If the temperature difference across the heat exchangers is 5°, the heat transfer capability is 1,500 to 2,250 Btu per hour per square foot, and the 340,000 Btu per hour of heat transfer equates to 225 to 150 square feet of heat exchangers for seawater application for each kW of rating. If pessimistic assumptions are made concerning each of the three factors--the heat transfer coefficient of heat exchangers in marine service, the allowable temperature drop across the heat exchangers, and the cost per square foot of heat exchangers for marine service--then the cost of OTEC can seem prohibitively high. On the other hand, if optimistic assumptions are made concerning each of these three factors, then OTEC seems competitive even at present costs of electricity.

It is unfortunate that a number of years have gone by, and significant effort has been spent on design studies for OTEC plants, without these important uncertainties about OTEC heat exchangers being resolved. It is clear that disputes about these uncertainties will not be quelled without actual measurements performed in the marine environment. Fortunately, the budget for the OTEC R&D program has recently been increased to the point that laboratory and field experiments can now be supported. These experiments should provide sufficient operating data to resolve much of the uncertainty. Tests of marine heat exchangers for OTEC applications are now a high priority item in the government solar energy program. Research is also under way on advanced designs for heat exchangers based on plastic or other low-cost materials that may lead to significant reductions in heat exchanger costs. There are some who are very hopeful this research will be successful; others are very skeptical.

The various estimates of the cost of ocean thermal plants seem at first to show a wide variance. However, at least part of the variance can be traced to identifiable assumptions about the plant and its site. Table IV.14 shows the estimates of several studies. It is seen that correcting the estimates for the effects of different temperature drops between the warm and cold water sources and correcting for the differences in the assumptions about the costs of marine service heat exchangers brings the estimates into reasonable agreement. The costs in this table do not include the cost of transmission to shore.

The cost of transmission to shore is also uncertain. For underwater transmission of electricity for long distances, direct current is required. Thus ac/dc converter stations will be needed. One estimate puts the cost of undersea transmission over a distance of 200 miles (depth of 2,000 feet) at about 13 to 15 mills per kWh (Arthur D. Little, 1975). This has the effect of adding the equivalent of about $800 per kW to the capital cost of the OTEC plant. The cost of transmission for a particular OTEC plant will depend strongly upon local conditions.

If the construction and operating costs of OTEC plants are indeed low enough for there to an economic incentive, then it seems their deployment can be fairly rapid. All the materials required by an OTEC plant are commercially available in the U.S. and are abundant throughout the world: steel, concrete, copper, aluminum, and even titanium. The OTEC plants can be built in shipyards similar to those in existence in other countries. The North Sea oil technology is more advanced and elaborate than that demanded by OTEC plants. In addition, OTEC plants can be replicated, much like building a family of large ships. Thus, there is no need for the time-consuming reengineering associated with conventional plants constructed on land.

Our estimates of the energy contributions of OTEC plants are based on the assumption that they are found to be economically competitive, and that they are, therefore, deployed rapidly to meet the growth in electrical energy demands for the Gulf States (Table IV.15). The cost estimates are not intended to include the costs of the long-distance

TABLE IV.14 Comparison of Several OTEC Plant Cost Estimates

Characteristic	Lavi[a]	McGowan *et al.*[b]	Dugger *et al.*[c]	TRW[d]	Lockheed[e]
Heat Transfer (Btu/Hr/Ft^2/°F)	1,000	NA	~450	~450	~450
ΔT, °F	40	32	39	40	34
Working Fluid	NH_3	Propane	NH_3	NH_3	NH_3
$/kW Excluding Operating Cost	656	712	357	1,812	2,594
$/kW Excluding Hull, Cold Water Pipe, Mooring & Deployment	559	610	262	1,102	1,901
$/kW Adjusted to $\Delta T = 34$	839	524	370	1,654	1,901
$/kW Adjusted to ΔT = 34°F and $9/ft^2 Heat Exchangers	978	NA	1,540	1,654	1,901

[a]Lavi, Abrahim. 1975. Solar Sea Power Project, Final Report. Springfield, Va.: National Technical Information Service (PB-242264), pp. 108-113.
[b]McGowan, J. G., J. W. Connell, and R. Braren. 1974. Variations in Heat Exchanger Design for Ocean Thermal Difference Power Plants. Springfield, Va.: National Technical Information Service (NSF-RA/N-75-110), p. 5.
[c]Dugger, G. L., H. L. Olsen, W. B. Shippen, E. G. Francis, and W. H. Avery. 1975. Tropical Ocean Thermal Power Plants and Potential Products. New York: American Institute of Aeronautics and Astronautics (75-617), pp. 23-26.
[d]TRW. 1975. Ocean Thermal Energy Conversion, Research on an Engineering Evaluation and Test Program, Vol. 3, Baseline System Design. 5 vols. Springfield, Va.: National Technical Information Service (SAN-1089-T1 [part 3]), pp. 5-1 to 5-5.
[e]Lockheed Missile and Space Company. 1975. Ocean Thermal Energy Conversion Power Plant Technical and Economic Feasibility, Vol. 1: Technical Report. Springfield, Va.: National Technical Information Service (NSF/RANN/SE/GI-C937/FR/75/1), sec. 1, pp. 87-89.

transmission that would be necessary for OTEC to meet electrical demands in other parts of the nation.

OTHER RENEWABLE ENERGY TECHNOLOGIES

The Solar Resource group has considered briefly the potential for achieving energy contributions on a national scale before 2010 from other renewable sources, or from other technologies that use solar energy. Such alternatives include ocean waves, tidal power, and space satellites for solar energy conversion and transmission to Earth. We have decided that prospects warrant including these technologies in the national energy scenarios and economic models being used in the CONAES study. This is not to say that they may not provide some energy that will be important to some local applications, or that they might not become important in the far future. Thus, R&D on such alternative approaches is encouraged but their formal inclusion in the CONAES study is judged unnecessary.

TABLE IV.15 Potential Energy Contributions and Costs, Ocean Thermal

Level of Technology	1975	1985	1990	2000	2010
Energy Contribution, Quads[a]	--	0.01	0.08	1.6	3.9
Rated Capacity, MWe	--	100	1,000	20,000	50,000
Load Factor	0.9	0.9	0.9	0.9	0.9
Present Technology[b]					
Cost of OTEC Plant, $/Rated kW[c]	5,000	2,600	2,600	2,600	2,600
Cost of Transmission System, $/Rated kW[d]	2,000	800	800	800	800
Cost of Electricity, Mills/kWh[e]	137	70	70	70	70
Advanced OTEC Technology[f]					
Cost of OTEC Plant, $/Rated kW	--	1,990	1,250	1,100	1,100
Cost of Transmission System, $/Rated kW[g]	--	800	600	400	400
Cost of Electricity, Mills/kWh[e]	--	57	39	33	33

[a]Converted from kWh at 10,000 Btu per kWh.

[b]Corresponds to Lockheed design with titanium heat exchangers with heat transfer coefficient of 450 Btu/hr/ft^2/°F.

[c]Lockheed cost estimate assumed constant from 1985 on.

[d]Based on an example of 200 miles to shore with high voltage dc transmission and ac/dc conversion at both ends. Actual cost will depend strongly on site.

[e]Based on 15 percent charge on capital and $33 per rated kW annual O&M cost. This is cost of electricity delivered to shore.

[f]Assumes successful development of an advanced heat exchanger (perhaps aluminum) of low cost but with a heat transfer coefficient of 1,000 Btu/hr/ft^2/°F.

[g]Assumes development of an advanced technology for undersea power transmission with low costs and high reliability.

V. COMMENTARY ON THE SOLAR ENERGY R&D PROGRAM

THE LEVEL OF EFFORT

Federal funding for solar energy R&D has grown rapidly since the beginning of this decade. The growth is shown in Figure V.1 and detailed in Table V.1. This growth has partially been motivated by the perception of possible economic and environmental attractions of solar energy, but it has also been spurred by a sense of urgency that other energy sources may soon be exhausted, or be withheld for political reasons, or be too dangerous.

Those who feel this urgency most strongly have generally turned to solar energy as the ultimate alternative. To those with this sense of urgency it seems valid to make comparisons with efforts at obtaining a nuclear weapon in the early 1940s and in getting to the moon in the 1960s. These efforts were based not on economic comparison with alternatives, but rather on questions of urgency. The question is then raised whether the solar program should be in the same category.

The present federal program is planned to achieve measured progress at minimum risk. When there are several alternative solutions to a technical problem, the solutions are generally tried in sequence of perceived feasibility rather than being tried in parallel. Thus a solar thermal pilot plant is planned for the central receiver design with a Rankine cycle, but the government has not decided to proceed in parallel with pilot plants of other concepts for solar thermal plants. The element of the federal budgetary process that has been the most important advocate for the measured pace of the government solar energy program is the Office of Management and Budget, which is also probably the single federal organization that is most negative in its perception about the prospects for widespread use of solar energy.

The decision to have a solar energy program, even of measured progress, is largely due to a general concern about the eventual depletion of fossil fuels. This concern is not generally perceived as urgent despite the fact that known reserves of oil and gas will last less than 10 years. Although exploration will undoubtedly add to these reserves, it is not at all clear that the government projections of new production will be met. This is a first reason to raise the question of urgency, since the present energy supply system is so massive as to require decades for conversion.

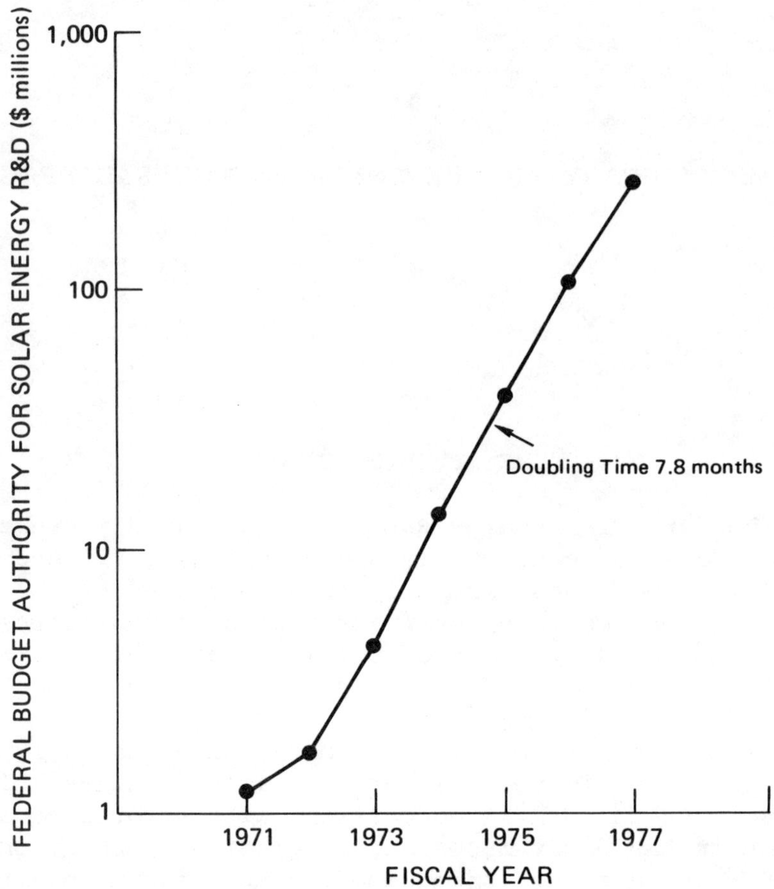

FIGURE V.1 Federal budget authority for solar energy R&D.

A short-term conversion to imported oil and gas has been assumed by many policy analysts, but the propriety of massive dependence on imports is questionable. Our huge national coal resource could also be used by conversion techniques to meet our needs for oil and gas. However, this option may be limited by the availability of water in the western states, or by standards restricting pollutant emission in pristine areas, or ultimately by a worldwide moratorium on combustion, to avoid a global greenhouse effect. A conversion to nuclear sources has also been generally assumed, at least until recently. Now questions of risk, siting, waste disposal and uranium availability are making the public acceptance of this alternative less sure.

Thus, the solar alternative appears to some as the only one now free of major noneconomic constraints. It is argued that so little has been spent on R&D for solar energy compared to other energy technologies, and the range of solar alternatives is so large, that there seems little question that cost reductions will follow from increased federal expenditures. The question then is one of the degree of urgency: whether alternatives will be funded simultaneously rather than sequentially. The decision is appropriately a political one, and one which can ultimately

TABLE V.1 Federal Budget Authority for Solar Energy R&D, in Millions of Dollars

	Fiscal Year							
Solar Type	71	72	73	74	75	76	77[a]	78[b]
Heating and Cooling	0.5	0.1	0.8	5.8	12.0	33.6	84.8	44.8
Process Heat	0	0	0	0	0.5	4.7	7.8	10.3
Bioconversion	0.6	0.4	1.0	1.2	0.6	5.2	9.7	17.0
Thermal Electric	0.1	0.6	1.4	2.7	13.2	14.3	51.3	121.6
Photovoltaic	0	0.3	0.7	2.7	5.0	21.6	59.4	50.5
Wind Energy	0	0	0.2	0.8	5.3	14.9	20.5	25.7
Ocean Thermal	0	0.1	0.2	0.6	1.9	8.1	13.5	26.0
Support	0	0	0	0	1.4	6.2	11.5	9.0
TOTAL	1.2	1.5	4.3	13.8	39.9	108.6	258.5	304.9

[a] Tentative budget, preliminary to passage of fiscal year 1977 authorization bill.

[b] Budget presented by President Ford, January 1977.

be decided only in the context of perceived national urgency. There is now little perception of urgency, but the question must be raised whether that perception is correct.

It is expected that most decisions on future federal funding for solar programs, as well as individual and corporate decisions on particular solar systems, will be based on some version of life-cycle-cost economics. The decision is basically one of trading off future costs of capital for future costs of fuel. Since few external costs enter into these decisions and since we value the present much more than the future, most near-term decisions will favor nonsolar approaches. We point out here some of these additional costs and benefits which do not appear in traditional economics; mechanisms for their inclusion in estimates of cost are, however, beyond the scope of this report.

Environmental Benefits

It is generally perceived that most solar energy systems are essentially benign. If decentralized and properly handled, the two main potential disadvantages of extensive land use and adverse impact on aesthetics can be minimized or avoided. Because the environmental costs of competitive nonsolar systems do not generally appear in economic analysis, there is little present acknowledgment of this solar advantage, even in national decisionmaking.

Future Generations

It is clear that our present oil and gas resources are finite, rapidly declining, and more valuable for future chemical feedstocks than for their thermal energy content. Because of high interest rates (and, therefore, of high discounting of future benefits), it is not in a corporation's or individual's best economic interest to forgo present consumption of these commodities. Because there is no national tradition of husbanding resources, we continue to exploit them. Some solar energy applications would be more nearly cost competitive if a financial value were placed on the conservation of hydrocarbons; present economics do not give any credit, even in national economics, to this beneficial aspect of solar systems.

World Posture

With a solar-based economy, our domestic consumption of energy would not be occurring at the expense of resources also sought after by the rest of the world. With our present and projected energy consumption rates, it is clear that questions of international ethics and equity will be raised in the future as other nations become unable to compete economically for the limited world supply of exhaustible energy resources. Present economic comparisons do not include this value of using inexhaustible domestic energy resources.

It is important that the above considerations, as well as direct comparisons of life-cycle costs, enter into decisions as to the appropriate level of effort for the solar energy R&D program.

STAFFING AND ORGANIZATION

The present (Fall 1976) staff of the ERDA Division of Solar Energy (DSE) is approximately 100; fewer than 60 are professionals whose time is devoted to the technical tasks of planning the solar program, evaluating proposals, and monitoring contracts. It has been reported in Congressional testimony that a majority of all ERDA's mail deals with solar energy and that Congressional demands on DSE staff time are very high. There is a general perception that the DSE staff is badly overworked so monitoring of programs has suffered, and the DSE staff is often unable to make site visits to projects because of a chronic shortage of travel funds. It appears likely that the taxpayer's best interests are being ill-served by false economies. The Office of Management and Budget (OMB) has not granted requests for increased staffing. Because the division personnel requests are not considered part of the budget process, they do not appear in the Holifield tables, and the Congressional committees do not include detailed questions on staffing as part of their authorization or appropriation process.

One important problem of the organization of the solar program is that all advanced energy storage research is now handled within the Office of Energy Conservation. An ERDA-approved FY77 request to OMB for

$10 million for solar storage research within the solar division was disapproved without a corresponding increase for the Office of Energy Conservation. The rationale for disapproval was, of course, to avoid redundancy and to promote synergisms with other energy storage programs. However, whereas improved energy storage is important for other energy technologies, it is critical to the success of some of the elements of the solar program. It is also clear that certain small-scale energy storage research programs are unique to solar energy applications.

RECOMMENDATIONS FOR SOLAR ENERGY R&D

The question of the direction and relative effort that ought to be expended on the various elements of the solar energy R&D program has been extensively debated. It has been found impossible to arrive at a consensus on this question among those active in solar energy, largely because of the variety of the solar technologies, their different states of development, the impossibility of judging the long-term benefits of the technologies still in the stage of advanced research, and the conflicting interests of those who foresee themselves being best served by different types of technologies. There has been no consensus within the Solar Resource Group on all aspects of the question of direction and relative emphasis within the solar R&D program. However, there has been at least partial agreement on a few issues, and these issues are discussed here. Many of the points made here are also presented in the corresponding parts of Chapter IV. Issues regarding implementation are discussed in Chapter VI.

Heating and Cooling

The federal solar heating and cooling program has thus far emphasized the development and demonstration of active space heating systems. This is an advanced technology approach to the problem of providing space heating. It is perhaps appropriate that NASA has been called upon to provide technical support to the heating and cooling demonstration program, since the style of this program is compatible with that agency's considerable skills of technology development through closely defined goals and strong management. The goal of this program is to develop and demonstrate the adequacy of a technology package that can serve as a direct replacement for conventional heating systems, including all normal features such as thermostatic control. Such systems are intended to be compatible with fairly standard design and construction practice, and in operation to be transparent to the occupant. This program may indeed be successful in establishing a solar heating industry, but there is an alternative approach to solar heating that has thus far been generally neglected in the government program.

The alternative approach might start with a different statement of the goal of the heating and cooling program: to learn to design, construct, and operate buildings that use to best advantage the features of external environment, especially sunlight, to provide a comfortable and pleasing living environment. This statement properly puts a heavy emphasis on architecture, for it is in architecture that the federal program has fared most poorly. The government's demonstration homes tend to be standard, tract-style houses with a high-technology solar heating system tacked on. Conservation is served by adding some insulation, and perhaps using double-paned windows. Ignored is the question of altering the fundamental design of the house to make it more compatible with solar energy.

This alternative concept of the goal of solar heating and cooling is the motivation for the design of passive solar energy systems. A number of successful passive designs have now been demonstrated. Some of these have been highly imaginative, some have not. Quantitative measurements of their performance are largely lacking. However, it is now clear that this approach has significant promise. The advances made thus far in passive system design have taken place almost entirely outside the ERDA/HUD/NASA program. The government should now recognize the importance of this approach, and should assign a high priority to supporting imaginative work in this area. Some of the things that need to be done include supporting architectural research, including support for the students who will be the next generation of architects; establishing better methods for analysis of the performance of such systems, including correct treatment of the processes of infrared heat transfer and natural convection; determining the performance of various designs as influenced by local climate; assisting in the development of building standards that allow for the use of passive design concepts; and providing support for the construction and monitoring of experimental designs, especially those of unusual construction.

Process Heat

The apportionment of government solar R&D funding does not seem to reflect the significance of the potential contribution of solar energy to industrial process heat production. About 25 percent of total U.S. energy consumption is used to produce heat for industrial processes. Between half and two-thirds of this heat is in the form of hot water or low-temperature steam (<350°F). This is about equal to the space-heating requirements of all residential and commercial buildings. Yet the ERDA budget for FY77 provided less than 1/10 as much funding for process heat as for heating and cooling of buildings.

It seems possible--with the necessary research, demonstration, and development and tax incentive support from the federal government--for solar energy to supply at least 5 percent of the energy unused in industry for making hot water and low-temperature steam within the next 10 years. There are several reasons why industry may accept direct use of solar energy more readily than homeowners. Life-cycle costing is the accepted way that industry evaluates an alternative energy system, whereas the large capital cost of a residential solar system is a major deterrent for the developer or owner of a new home. About 50 percent of industrial process heat now comes from natural gas. Industrial users of natural gas are being placed on interruptible service and in some areas gas supply to industry is being terminated. This provides a natural incentive for industry to look to new energy sources--including solar. A solar process heat system can provide heat year-round rather than just during the winter months. Hence, the annual usable heat output per unit collector area is about twice that of a residential solar heating system. Moderate operation and maintenance costs are acceptable for process heat systems. Maintenance personnel are always available.

There is a strong implication in government planning documents that progress in solar technology for process heat is dependent on progress in heating and cooling technology for residential buildings. This is incorrect because different types of collectors will be required to provide process heat. Since large area systems will be required for process heat, this means that cost savings can be obtained by designing large area, low-cost collectors with minimal requirements for intercollector manifolding and plumbing. Significant use of solar process heat will depend upon the development of low-cost collectors to provide low-pressure steam. Such collectors are not of high priority to the space heating and cooling program.

The government should accelerate its program on solar process heat, and give high priority to developing collectors and storage systems that are compatible with standard industrial steam systems.

Bioconversion

The question has been raised repeatedly, by the Office of Technology Assessment and others, whether the government program overemphasizes the generation of electricity in comparison with the production of storable and transportable fuels. It is clear that many elements of our energy system, especially transportation, can only with difficulty, and perhaps at severe economic disadvantage, convert to use of electricity instead of liquid or gaseous fuels. Since it is liquid and gaseous fuels that are now in shortest supply, and since both coal and nuclear can supply electricity, it would seem appropriate to assign a high priority in the solar energy program to developing technologies that could provide, at good efficiency, liquid or gaseous fuels from solar energy. However, this is now an area of relatively low priority. Table V.1 shows that the section of the government program with the major responsibility for production of fuels, the bioconversion program, has lagged far behind the four solar electric technologies in level of funding. In FY77 the planned allocation is $144.6 million for solar electric, and $9.7 million for bioconversion.

There are actually a variety of possible techniques for producing fuels from solar energy in addition to the concepts now being supported in the bioconversion program. Alternative concepts include high-temperature cycles for production of hydrogen, electrolysis of water, and direct photochemical production of fuels. If solar energy is to be able to provide storable and transportable fuels, then the areas already included in the bioconversion program deserve a higher level of support than has been given thus far, and these alternatives should receive priority attention.

The area of solar energy R&D that might have the greatest impact in the long term is direct photochemical production of fuels. There are a variety of ideas as to how this might be done (see, for example, Porter and Archer [1976]). However, this is an area of advanced research of the type that the present ERDA solar energy program has not adequately supported. A dedicated long-term effort on direct photochemical production of fuels seems amply justified by the potential impact of a successful development in this field.

Solar Thermal Conversion

The major question in the solar thermal conversion program concerns the allocation of significant expenditures in this area between central station concepts, such as the central receiver, and distributed generation concepts, such as solar total energy systems. It is clear that government efforts are increasing on distributed electrical approaches, which had previously received very little funding. A budgetary line item appeared for solar total energy in the FY77 funding for the first time. However, the program for developing distributed electrical systems is still not well defined, having been previously dominated by heavy emphasis on the central receiver concept for central station generation. In particular, there is virtually no planned development of small prime movers suitable for distributed systems.

There have been no government studies comparing the economic advantage of centralized versus decentralized electrical generation, though this has become a hotly debated subject.

A proponent of distributed solar electric systems would make the following points: distributed systems have the advantage of being more amenable to assembly-line economics and subsequent rapid installation. If backup power is generated locally or sufficient storage is available, the costs of transmission and distribution systems can be eliminated. The use of waste thermal energy in total energy systems is possible only with distributed systems. Lastly, a failure of one unit is much less likely to cause failure of the entire system.

A proponent of centralized systems in response would argue that no small prime movers have been shown to be economical compared to large units; that adequate storage technology does not exist and the cost of providing backup power is prohibitive because of the poor peak-to-average load ratio it imposes on the utility; that the cost of marketing small units causes an intolerable mark-up; and that individuals and small companies do not have adequate access to capital to finance such systems.

High priority objectives for the federal solar thermal program should be to clarify the pros and cons of distributed and centralized solar electric systems, and to assure that the R&D program is appropriately balanced between these approaches.

Photovoltaic Conversion

As was discussed at length in the photovoltaic part of Chapter IV, the present government photovoltaic program puts insufficient emphasis on advanced research on photovoltaic materials and processes. Such research will be required if any photovoltaic technology is to meet the long-term goals of the program.

Wind Energy

As the wind energy part of Chapter IV proposed, the government should let contracts for a dozen or so 1-MW units so as to get the design effort into firms accustomed to cost-effective design and production; establish mock clusters of anemometers to determine the relationship of resource to utility loads; and develop plans to subsidize the initial development of a wind energy industry.

Ocean Thermal

The highest priority in the ocean thermal conversion program should be to settle, by field experiment, the question of the performance of OTEC heat exchangers in their proposed operating environment. This research is now under way in the government program.

VI. COMMENTARY ON IMPLEMENTATION OF SOLAR ENERGY

The problem of achieving a rapid and efficient implementation of new energy technologies has received great attention in ERDA planning documents. It has also been the subject of several commentaries on the ERDA program by the Office of Technology Assessment, the Federal Trade Commission, and others. Despite this extensive discussion, it is still unclear just how new technologies might enter the energy market, which actions of the federal government will aid that entrance and assure competition in the new energy industries, and which federal actions will actually impede the adoption of new technologies. These questions are probably most difficult and most pressing in the context of energy conservation technologies, since these are tied so intimately to the many sectors of final demand and their diverse institutional arrangements. However, these questions are also of great importance to the development and use of solar energy technologies.

COMPETITION OF ENERGY SOURCES

In analyzing the economic feasibility of a solar energy technology, it is typically assumed that if the solar option turns out to be more expensive--for example by 25 percent--than an alternative, then that solar option is not economically competitive and will not gain a market. However, such an assumption is in contradiction with present experience in the energy market. A wide variety of forms of delivered energy share the total energy market in a manner that is not always obvious from simple analysis of cost to consumers. The most frequently chosen type of space heating for new housing is electric resistance heat. This is true both in regions where no new contracts for natural gas or fuel oil are available (and the choice of electrical resistance has an obvious rationale), and also in regions where these alternatives are available. Where available, natural gas is about five times less expensive than electrical resistance heat, and fuel oil, even at present prices, is about half the price of electricity. Why then is the seemingly more expensive option chosen so often? Why is first cost an overriding factor? Who actually makes the choice? These questions must be answered if we are to understand the economic considerations that will be applied by consumers in deciding whether or not to install solar heating or cooling in their new homes.

The above is only one example of how strange the present structure of the energy market seems from an examination of only immediate costs to consumers. If the consumer's actual criteria of choice (including the criteria of utility companies that are consumers of electrical generating equipment) are understood, we will be able to predict whether particular solar applications will find a share of the market.

RETROFIT OF BUILDINGS FOR SOLAR ENERGY

It is well known that this nation's huge inventory of existing buildings is a major obstacle to the success of any effort to significantly reduce energy consumption. Whereas energy-inefficient appliances and transportation vehicles are quickly replaced, buildings are replaced only slowly and many of our present buildings will continue to exist in the year 2010. Unfortunately, the little effort that has been made at any level of government to reduce energy consumption in existing buildings has been without much attention to the possible role of solar energy in existing buildings.

There is no doubt that the major barrier to retrofit is economics. It costs more to retrofit than to design for solar energy. But another barrier is education. Broad experience both in the use of solar energy and in the design and construction of buildings is necessary in order to intelligently (simply and economically) synthesize solar energy with existing buildings. Unfortunately, buildings vary so greatly from each other that there are literally millions of different conditions and situations. There is virtually no incentive at the present time for any one group of people in the private sector to make the necessary (and huge) analytical and educational effort.

IMPACT ON ELECTRIC UTILITIES

Electric utilities are really energy utilities--energy is their business. Their prime duty is to continue to provide electric energy for society in the most reliable and economic manner, taking due cognizance of all potential environmental impacts and taking all due measures to ensure that these are minimized. Thus, as far as solar energy is concerned, the electric utilities' goal should be to aid in its development in a manner which will maximize its benefits to all members of society and minimize its undesirable impacts. It is not in the best interests of society if solar energy applications are developed in an uncontrolled manner without regard to their effect on other energy sources and their relative cost.

To appreciate the impact that the development of solar energy applications might have on electric utilities, it is useful to examine the load requirements which utilities must meet on a daily and on a seasonal basis. For example, Figure VI.1 shows a typical load curve for a medium-sized utility in the Northeast for a day in January and a day in June. In fact, these are for Boston Edison Company, but they are sufficiently generic to be applicable to other utilities as well. Figure VI.2, which

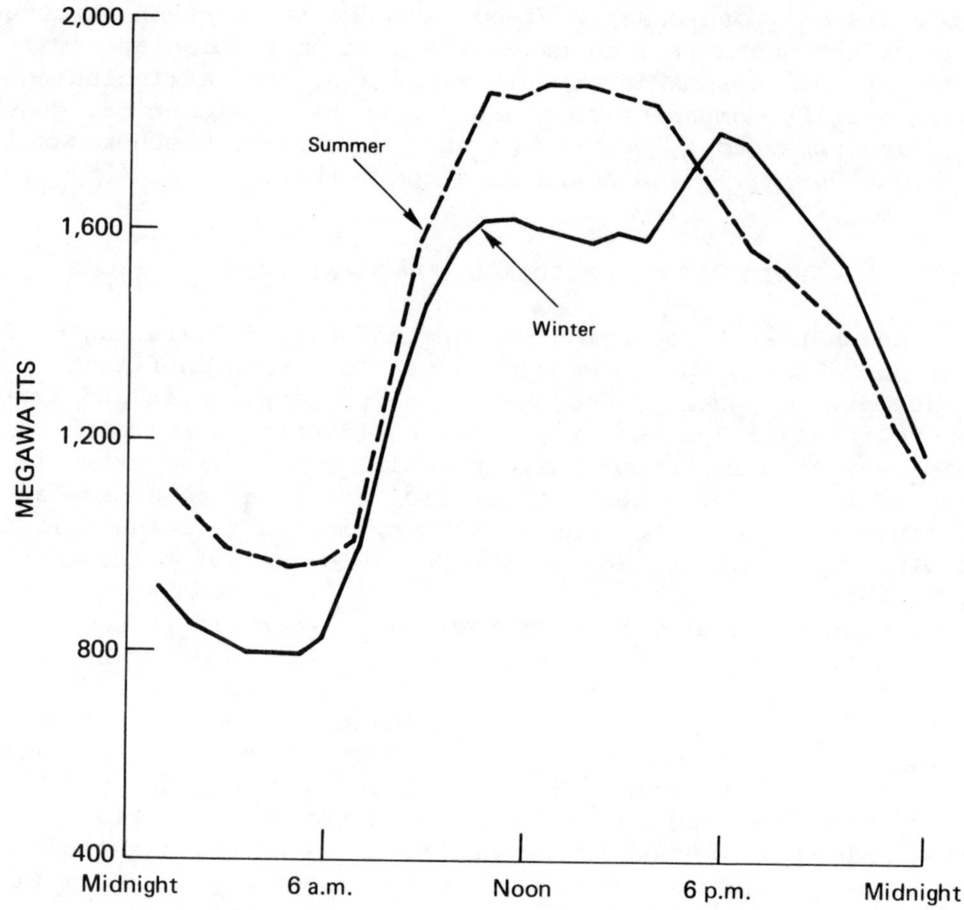

FIGURE VI.1 Daily load curves for a typical northeastern electric utility.

shows this daily load plotted on a weekly time frame, shows that:

1. the pattern is repeated;
2. there are three different types of load, namely,
   base load--always required,
   intermediate load--usually required more than six hours per day, and
   peak load--required less than six hours per day.

The base load generators (large fossil and nuclear) are the most economic; the next are the intermediate; and the least efficient and most expensive are the peak load generators. It is also instructive to replot these curves, as in Figure VI.3, to show what fraction of the total yearly load is peak, intermediate, and base.

Utility operations would benefit if the peak load could be transferred as much as possible to base load, or failing that, to intermediate load. Such transfers would save fuel or energy (because peak load generators are least efficient) and thus help reduce or keep down the cost of

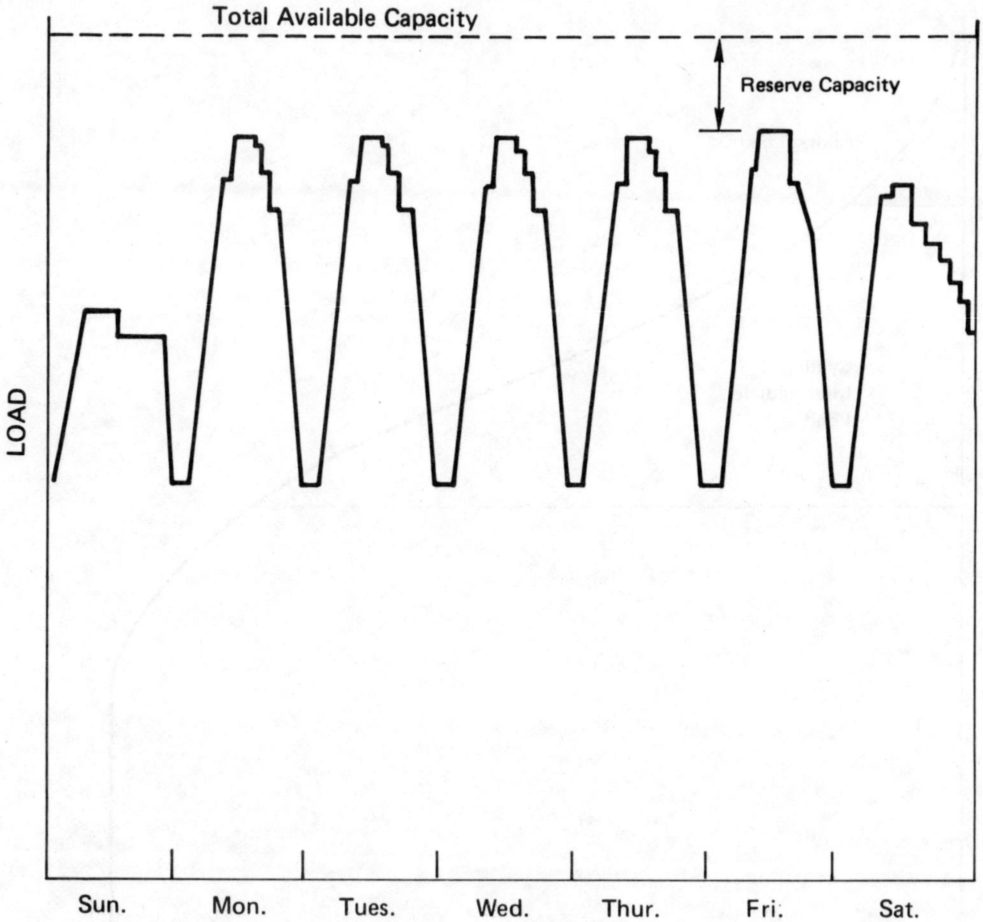

FIGURE VI.2 Weekly load curve for a typical northeastern electric utility.

electricity, while at the same time making this fuel available for other needs of society. There is yet another important benefit to be derived from reducing the peak load, which is the potential for capacity saving. That is, the utilities may be able to reduce the amount of generation capacity either by retiring less efficient old units or by deferring the installation of new capacity. This latter can represent significant capital savings, which of course could be reflected in the cost of electricity. All the above constitute a favorable impact on the electric utilities and can be achieved, for example, by using energy storage or some form of load management or control.

To remove part of the base or intermediate loads without reducing the peak load (or only slightly reducing it) would not be desirable. This would result overall in less efficient generation of electricity, by using more fuel on the average per kWh generated, and would not allow for any savings in capital costs. The net effect would be to increase the cost of electricity for both the utility and the customer, clearly an unfavorable impact.

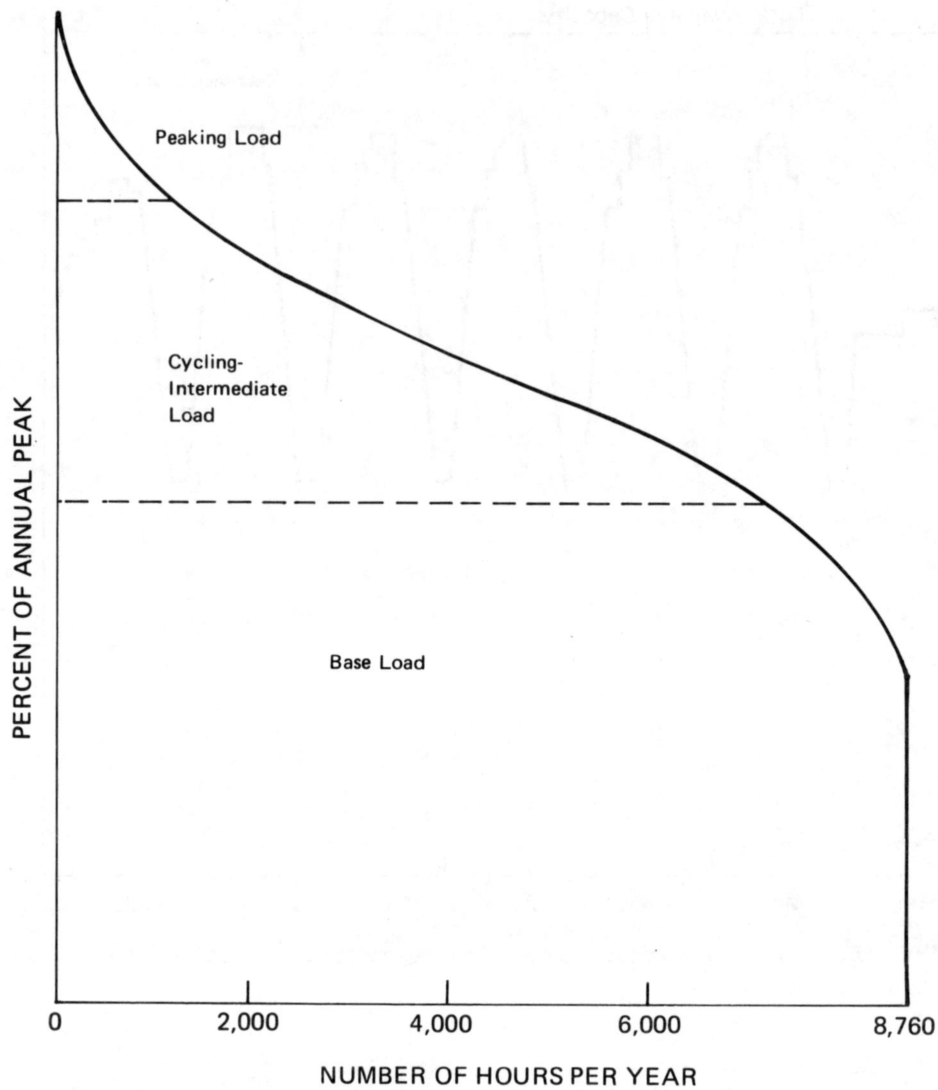

FIGURE VI.3 Duration of system load as a function of percentage of system peak load.

It is important to determine whether use of solar energy is likely to have a favorable or unfavorable impact on the electric peak load and, if the latter, what, if anything, can be done to change it to a favorable impact. Before attempting to answer this question it is necessary to determine when and how much solar energy is available. This will vary in different parts of the U.S. However, the generic shape of the plots of solar insolation versus time of day will be similar to those shown in Figure VI.4 for a clear day in June and a clear day in January in the Boston area. For the winter day, no solar energy is available before 7:30 a.m. and after 4:30 p.m.; for the summer day this "solar window" is extended by about two hours on either side. The shape of these curves on any given day will be affected by the presence of clouds.

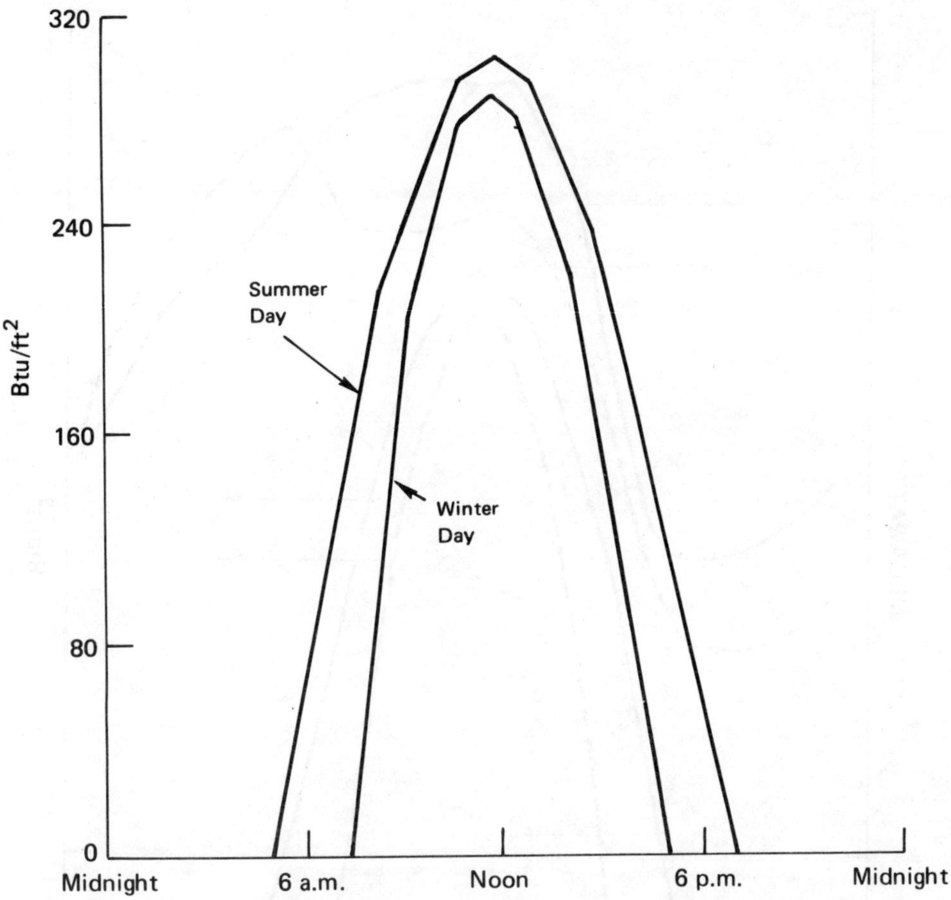

FIGURE VI.4 Daily distribution of insolation in Boston for clear summer and winter days.

The potential impact of solar energy on electric utilities can be examined by superimposing plots of solar energy availability on plots of utility daily load duration curves for the various seasons. This is done in Figure VI.5. It should be noted that the solar curves show only clear day solar energy availability per unit area rather than total actual usable solar energy for the various conversion applications. Thus, only general indicators or guidelines, rather than conclusions, are warranted. They are as follows:

1. During winter, use of solar energy is likely to have an unfavorable impact on the utility load factor by removing part of intermediate load with little or no effect on peak load.
2. During summer, solar energy can have a favorable impact on the utility load factor *provided* that useful applications can be found for solar energy during hot summer days. There are serious difficulties to achieving this, not the least of which is the lack of availability of suitable solar cooling systems and the low duty

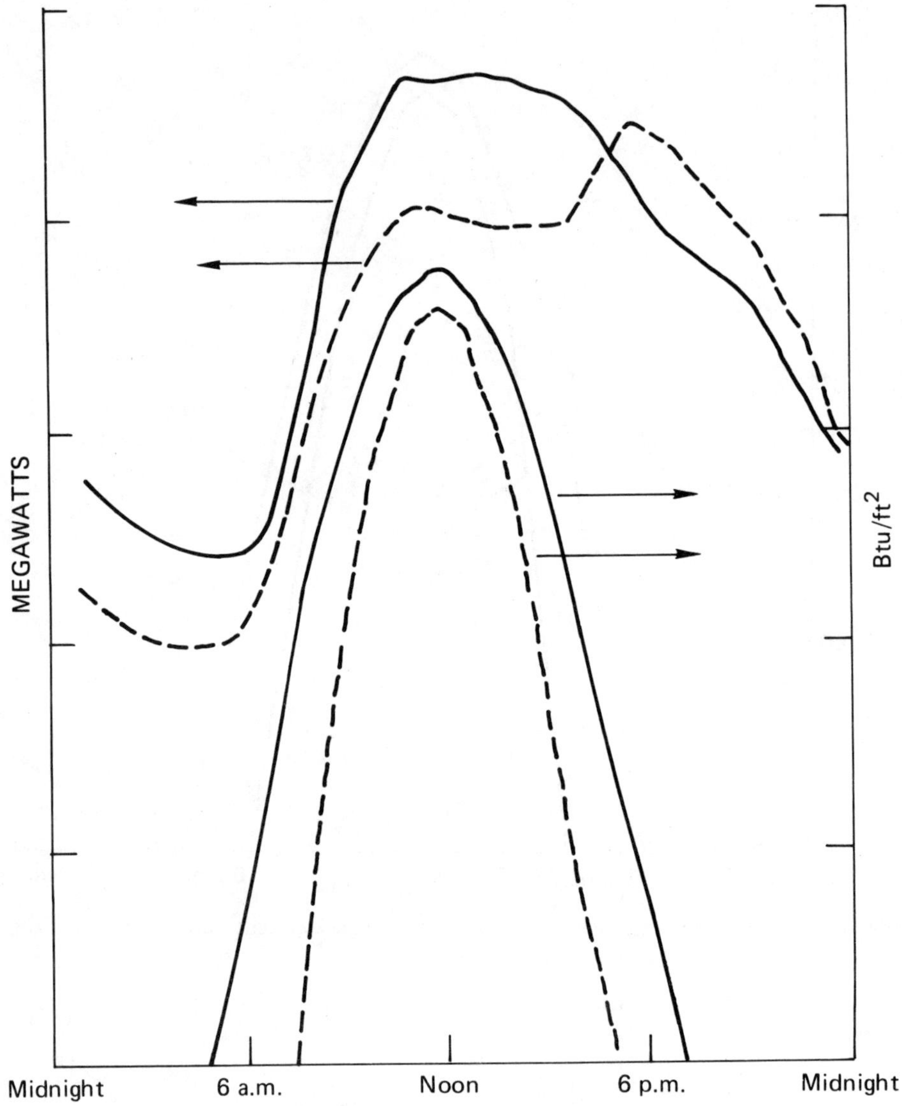

FIGURE VI.5 Comparisons of the daily distribution of clear day insolation and daily load for a northeastern electric utility. NOTE: Summer, solid line; winter, dotted line.

factor of residential cooling units. Solar energy is already responsible for a significant part of the peak load in summertime, by itself creating the need for air conditioning.

3. If adequate storage and appropriate control systems were available, then the impact of solar energy in winter could be changed to favorable and the summer impact could also be improved. Load management and control strategies must ensure that the peaks are indeed shaved and the stored solar energy is used during peak periods.
4. If storage is available or provided, then use of solar energy can result in plant or capital savings as well as energy savings. Further, if storage (thermal or electric) is provided at the individual

load centers, as may be appropriate for some forms of solar energy, then there will be potential for effecting capital savings at the distribution level as well as at the generation and transmission levels.

We now consider the potential impacts on utilities of the various solar energy applications.

Solar Heating and Cooling

Currently, the heating and cooling of buildings and the provision of hot water rely on either electricity or fossil fuels, primarily oil and natural gas. However, the latter two are expected to be phased out as reserves diminish in the years ahead, with the gap to be filled by electricity, coal-derived fuels, or new sources of energy such as solar.

As shown in Figure VI.5, solar heating clearly has the potential for energy saving during the winter but, without storage and control systems might well have an undesirable effect on the load curve through reduction in intermediate or base load with little, if any, reduction in the peak load. With adequate storage, however, it should be possible to move a fraction of the solar energy contribution to the time of peak load sufficient to produce an improved load factor. In this respect, it is noteworthy that solar water heating has more flexibility than solar space heating because of its built-in storage. For the summertime, solar water heating can only help because the solar space heating load will be close to zero.

Solar cooling is entirely different--it is available during the day, when it is most needed, and is absent at night. Thus, even without storage it has the potential to improve the utility load factor and to produce both energy and plant savings. Unfortunately, systems for its application are not as well developed as are solar heating systems.

Solar heating also has potential applications in industrial and agricultural areas for process heat and crop drying, respectively. In the "sun belt" states the overall effect on utility load curves could be beneficial. In other areas there will certainly be energy savings but the effect on the load curves may be good or bad depending on a number of local factors.

Solar Thermal Conversion

In solar thermal conversion the energy input occurs only during the day and the concentrating collectors require direct solar radiation--hence, the main application is in the sun belt. The main difference from solar heating and cooling is that the utilities will need to provide the storage required to optimize the benefit from central station solar thermal conversion. Distributed solar thermal systems may provide their own storage, which may take any one of the number of forms: pumping storage, hydroelectric, compressed air, thermal, electrochemical (battery), inertial (flywheel), chemical (hydrogen), or superconducting magnetic energy storage.

Photovoltaic Conversion

Here the sun's energy is converted directly into electricity using arrays of solar cells. Again, the electricity is produced by direct solar radiation. The main difference (as far as the end product of electricity is concerned) between this and solar thermal conversion is that the solar cell arrays may be distributed at individual load centers. Thus, to optimize its use and prevent undesirable impacts on utilities, there may be need for storage at both the utility level and at the individual load center.

Wind Energy

As far as the utilities are concerned there is probably more need for storage in conjunction with wind power than there is with solar thermal and photovoltaic power simply because it is, in general, a less predictable source. Otherwise the same comments apply to the type and amount of storage desirable to optimize the use of wind energy.

Biomass Conversion

There would be no impact as such on the electric utilities apart from possibly adapting or designing generating facilities to use the fuel produced.

Ocean Thermal Energy Conversion

Present proposals in this area envisage fuel such as hydrogen being produced onsite and piped or ferried to shore; alternatively, the electricity produced may be transmitted via cable to shore. In the former case there would be no impact as such on the electric utilities; apart from making fuel available. In the latter case, however, it would appear that something close to base load power would be generated and, hence, would be useful to the utilities, provided the technological, economical, and environmental problems can be solved.

LEGAL AND INSTITUTIONAL CONSTRAINTS

There has been extensive discussion over the past few years of the importance of legal and institutional constraints on the use of solar energy. Yet it is still unclear just how important such constraints might be in hampering the adoption of a practical and economically competitive solar technology. The true institutional constraints will probably come to light only as we gain experience with each of the many diverse applications of solar energy. The Solar Resource Group has, of course, no way of predicting what factors will eventually be most important, but two things seem worth brief discussion here: the institutional arrangements for dispersed electrical generation, and the legal status of rights to sunlight. See also the reports of the Institutional Constraints and Incentives Resource Group (1977) and the Decision Making Resource Group (1977).

There has been little institutional experience in the U.S. with dispersed generation capacity that is owned by a local consumer, but is also connected to utility networks, nor has there been experience with total energy systems that provide both electricity and heat energy to the user. In particular, in many regions of the nation, it is not legal for an owner of a generating plant to put electricity back onto the net. In many cases the consumer is allowed either to have no generating capacity and depend entirely upon the utility, or to be completely self-sufficient and not be hooked to the utility at all. The widespread use of solar total energy systems will require that arrangements be made for consumers and utilities to interface their systems and share costs in an equitable manner. This is necessary for the success of all types of total energy systems, including fossil-fired systems as well as solar total energy systems.

Landowners in the U.S. have an inherent right to insolation only from directly above their property. No right exists to receive light across adjacent land unless a formal agreement or other legal arrangement provides for it. Since much useful sunlight does not impinge on earth from directly overhead, most owners of solar energy systems would benefit from a right to receive lateral light. Guaranteeing access to sunlight surpasses in importance all other legal issues that govern use of solar energy systems.

Outright purchase of adjacent property, purchase of air rights over adjacent property, or purchase of an easement for light can provide legally protected access to lateral sunlight, but the transaction costs and the difficulties of negotiating satisfactory agreements between property owners limit the appeal of these options. No single remedy, including new land-use planning methods, (e.g., retrofitting urban structures and developing new residential subdivisions), satisfies all conditions. Further, reliance upon private agreements requires a multitude of successful transactions before solar energy systems can contribute significantly to the energy budget.

Local, state, and federal legislative bodies have considered a variety of proposals to overcome existing laws concerning access to sunlight. Some of these provide direct legal assurances, while others offer various incentives to subordinate governmental units to guarantee some access to light. Public intervention in the market place in effect redefines property rights and mutual obligations of landowners. It decreases or eliminates transaction costs and, with proper planning, could result in no aggregate decrease in property values if the limitations on property rights are offset by the correlative gains.

Not all of these legislative proposals are legally sound or politically practical. Some are unduly vague because they fail to define necessary terms adequately, and others are inadequate because they specify or imply unconstitutional deprivation of private property rights. It is not at all clear which level of government should take the lead in quaranteeing access to sunlight, but federal encouragement coupled with state and local implementation might strike the proper jurisdictional balance and allow the necessary flexibility for adaptation to differing sites and structures.

REFERENCES

Arthur D. Little. 1975. Unpublished data, Cambridge, Mass.

Caleb, Charles. 1976. Not Out of the Woods. Environment 18(7):17-20.

Candela, B. J., and A. J. Wiener. 1975. Lifestyle and Energy Consumption: Scenarios for the Years 1985, 2000 and 2025--Alternative Coal Futures in the American Economy. Task 5 of Issues Relative to the Development and Commercialization of a Coal-Derived Synthetic Liquids Industry. Prepared by the Hudson Institute for the U.S. Energy Research and Development Administration. Springfield, Va.: National Technical Information Service (FE-1752-1).

Center for Strategic and International Studies. 1973. Understanding the National Energy Dilemma. Washington, D.C.: Georgetown University.

Consumption, Location, and Occupational Patterns Resource Group. 1979. Energy Choices in a Democratic Society. Report to the Synthesis Panel, Committee on Nuclear and Alternative Energy Systems, National Research Council, Washington, D.C.

Coty, U., and M. Dubey. 1976. The High Potential of Wind as an Energy Source. Paper presented at 2d Annual Energy Symposium, Los Angeles Council of Scientists and Engineers. North Hollywood, Cal.: Western Periodicals Company.

Decision Making Resource Group. 1977. Report to the Synthesis Panel, Committee on Nuclear and Alternative Energy Systems, National Research Council, Washington, D.C.

Donovan, P. I., W. Woodward, W. R. Cherry, F. A. Morse, and L. O. Herwig. 1972. An Assessment of Solar Energy as a National Energy Resource. Prepared by the Department of Mechanical Engineering, University of Maryland, for the National Science Foundation and the National Aeronautics and Space Administration. Springfield, Va.: National Technical Information Service (PB-221659).

Dow Chemical Co. 1975. Evaluation of New Energy Sources for Process Heat. Prepared for the National Science Foundation. Midland, Mich.: Dow Chemical Co.

Dugger, G. L., H. L. Olsen, W. B. Shippen, E. G. Francis, and W. H. Avery. 1975. Tropical Ocean Thermal Power Plants and Potential Products. Prepared by the Applied Physics Laboratory, Johns Hopkins University. Presented at the joint American Institute of Aeronautics and Astronautics/American Astronomical Society Solar Energy for Earth Conference, Los Angeles. New York: American Institute of Aeronautics and Astronautics (75-617), pp. 23-26.

Federal Energy Administration. 1974. Project Independence Blueprint, Final Task Force Report: Solar Energy. Springfield, Va.: National Technical Information Service (PB-248507).

Ford Foundation. 1974. A Time to Choose: America's Energy Future. Final Report by the Energy Policy Project of the Ford Foundation. Cambridge, Mass.: Ballinger.

Hewson, E. Wendell. 1975. Generation of Power from the Wind. Bulletin of the American Meteorological Society 56:660-675.

Institutional Constraints and Incentives Resource Group. 1977. The Supply/Delivery Systems for Energy Resources: Political, Legal, Regulatory, Institutional, and Financial Considerations. Report to the Supply and Delivery Panel, Committee on Nuclear and Alternative Energy Systems, National Research Council, Washington, D.C.

Kispert, R. G., S. E. Sadek, and D. L. Wise. 1976. An Evaluation of Methane Production from Solid Waste. Resource Recovery and Conservation 1:245-255.

Lavi, Abrahim. 1975. Solar Sea Power Project, Final Report. Prepared by Carnegie-Mellon University, Pittsburgh. Springfield, Va.: National Technical Information Service (PB-242264), pp. 108-113.

Lawrence Livermore Laboratory. 1976. LLL-Sohio Process Heat Project: Report No. 3. Lawrence, Cal.: Lawrence Livermore Laboratory (UCID-16630-3).

Lockheed Missile and Space Company. 1975. Ocean Thermal Energy Conversion Power Plant Technical and Economic Feasibility, Vol. 1: Technical Report. Springfield, Va.: National Technical Information Service (NSF/RANN/SE/GI-C937/FR/75/1), sec. 1, pp. 87-89.

McGowan, J. G., J. W. Connell, and R. Braren. 1974. Variations in Heat Exchanger Design for Ocean Thermal Difference Power Plants. Prepared by the Mechanical Engineering Department, University of Massachusetts. Springfield, Va.: National Technical Information Service (NSF-RA/N-75-110), p. 5.

MITRE Corporation. 1973. Systems Analysis of Solar Energy Programs. One of four MITRE reports on the National Science Foundation Five Year Solar Energy Research Program. McLean, Va.: MITRE Corporation (MTR-6513).

National Aeronautics and Space Administration and American Society of Electrical Engineers System Design Summer Faculty Program. 1973. TERRASTAR: Terrestrial Applications of Solar Technology and Research, Final Report. Prepared at the School of Engineering, Auburn University, Auburn, Ala. Springfield, Va.: National Technical Information Service (N-74-12674).

National Research Council. 1978. Energy Modeling for an Uncertain Future. Modeling Resource Group, Synthesis Panel, Committee on Nuclear and Alternative Energy Systems. Supporting Paper 2. Washington, D.C.: National Academy of Sciences.

National Research Council. 1979. Alternative Energy Demand Futures to 2010. Report of the Demand and Conservation Panel to the Committee on Nuclear and Alternative Energy Systems. Washington, D.C.: National Academy of Sciences.

Porter, George, and Mary D. Archer. 1976. *In Vitro* Photosynthesis. Interdisciplinary Science Reviews 1:119-143.

Reed, J. W. 1974. Wind Power Climatology. Weatherwise 27:236-242.

Teller, E. 1975. Energy: A Plan for Action. Report to the Energy Panel of The Commission on Critical Choices for Americans. New York: The Third Century Corporation.

TRW. 1975. Ocean Thermal Energy Conversion, Research on an Engineering Evaluation and Test Program, Vol. 3: Baseline System Design. 5 vols. Prepared by the Systems Group, TRW, for the U.S. Energy Research and Development Administration's Solar Energy System Final Report. Springfield, Va.: National Technical Information Service (SAN-1089-T1[part 3]), pp. 5-1 to 5-5.

Unger, Samuel G. 1975. Energy Utilization in the Leading Energy-Consuming Food Processing Industries. Food Technology 29 (December):33-46.

U.S. House of Representatives. 1974. Hearings on the Solar Energy Research, Development, and Demonstration Act, H.R. 10952. Subcommittee on Energy, Committee on Science and Astronautics. 93d Cong., 1st sess. Washington, D.C.: U.S. Government Printing Office.

U.S. Council on Environmental Quality. 1974. A National Energy Conservation Program: The Half and Half Plan. Washington, D.C.: U.S. Government Printing Office.

U.S. Energy Research and Development Administration. 1975a. A National Plan for Energy Research, Development and Demonstration. Washington, D.C.: U.S. Energy Research and Development Administration (ERDA 48).

U.S. Energy Research and Development Administration. 1975b. National Solar Energy Research, Development, and Demonstration Program: Definition Report. Washington, D.C.: U.S. Energy Research and Development Administration (ERDA 49).

Wilcox, Howard A., and Thomas M. Leese. 1976. Converting Seaweed to SNG. Hydrocarbon Processing 55(4):86-89.

